Originally published in 1916

A MODERN TWIN-CYLINDER MOTORCYCLE

A B C
OF THE MOTORCYCLE

Text and Illustrations that Make the Mechanism and Operation of the Machine Clear to those Directly or Indirectly Interested. A book for the Use of People who want the "SHOW HOW" Features.

By

W. J. JACKMAN, M. E.

Author of "Facts for Motorists," "Crushed Stone and Its Uses," and Similar Books.

ISBN: 978-1-940001-47-0

Sportsman's Vintage Press
2015

PREFACE.

The man who expects to find in this book a technical or theoretical discussion of the motorcycle and its mechanism will be disappointed. It has been written solely for the guidance of those who know they don't know much, if anything, about the subject, and wish to start right in advising themselves about it. With this purpose in view the language used is intentionally devoid of technicalities and readily understandable. The book is what may well be termed a "show how" one.

Step by step, in rational order, the various parts of the machine are taken up and their functions explained until these parts are assembled in the completed machine, and their unity of action thus demonstrated. First, we have the generation of the fuel which supplies the motive power; next, the manner in which the force created by this fuel is utilized; then, the part taken by the motor or engine in communicating this energy to the transmission system; and finally, the transmission itself. If the reader will give these various chapters careful consideration, comparing them as he goes along with the explanatory illustrations which accompany them, he will readily obtain a clear, comprehensive understanding, not only of the mechanism itself, but of the principle upon which it is based. This information, once thoroughly absorbed, will enable the rider of a motorcycle to get more satisfactory service from his machine, increase his pleasure in using it, avoid a lot of trouble in its operation, and enable him to overcome many annoying things which, under other circumstances, would require the attention of an expensive expert.

Contrary to general belief the principle of the motorcycle is simple. It does not require a technical mechanical training before it can be mastered. Any person of ordinary intelligence may understand it, and to give this understanding is the sole purpose of "A B C of the Motorcycle."

There is one chapter in particular—that treating of road troubles, their causes and remedies—which will be of great value to all users of this popular vehicle, whether experts or amateurs, and to it especial attention is invited. With those who give this feature of the book proper consideration "break-downs" will be less frequent and less expensive.

There is a fast-growing, profitable field of industry in motorcycle repairing which should be inviting to young men who desire to get into business for themselves. While not in any sense a textbook, "A B C of the Motorcycle" should appeal to young men of this class.

It is not the purpose of this work to advertise or exploit any particular make of motorcycle. In instances where the names of certain makes are used it is with the sole intention of illustrating, or making plain, some feature which is distinctive in the machine mentioned.

W. J. J.

The author desires to acknowledge the extension of many courtesies in the way of valuable advice and information, the loan of cuts, etc., by the following parties:

Mr. F. B. Hart, of the Excelsior Supply Co.; Mr. C. H. Lang, motorcycle expert; Mr. I. D. Whipple, "the motorcycle man"; Mr. T. H. Bullock, of Bullock Motor Co.; Mr. Edwin M. Snow, ignition expert; Mr. H. G. Hartenstein, Armac Motor Co.; Mr. George T. Briggs, Factory Sales Corporation; Mr. W. F. Hapgood, Hendee Mfg. Co.; Mr. Robert G. Elwell "Sparks"; Mr. R. D. Buell, president Chicago Motorcycle Club; Mr. T. J. Berger, editor Motorcycling; Mr. David Beecroft, editor Motor Age; Mr. W. C. Overman, vice-president Greyhound Motor Works; Mr. R. W. Green, Jeffrey-Dewitt Co.; Mr. J. F. Merkel, engineer Merkel-Light Motor Co.; Mr. E. P. Hovey, of Splitdorf Laboratory; Mr. W. A. Johnson, Pierce Cycle Co.; Mr. A. M. Dow, of Dow Mfg. Co.; Mr. R. B. Dunlap, of C. A. Shaler Co.; Mr. H. C. Genung, of Herring-Curtiss Co.; Mr. Harlow Hyde, of Bosch Magneto Co.; Mr. C. A. Mariam, of Reading Standard Co.; Mr. W. B. Harding, G. & J. Tire Co.; Mr. E. B. Smith, Veeder Mfg. Co.

CONTENTS

CHAPTER I.

INCEPTION AND EVOLUTION OF THE MOTORCYCLE.

In one form or another mechanically propelled vehicles of various kinds may be traced back to 1649. It is authentically known, however, that the first real motorcycle—a two-wheeled vehicle propelled by a gas engine—was constructed and operated by Gottlieb Daimler, a German engineer, in 1886. Previous to this, Otto, another German engineer, had produced a gas engine of

The First Motorcycle (Daimler's).

the stationary type, used exclusively for industrial purposes. In 1884, Daimler, who had been director of the Otto works at Deutz, designed a small gas engine, the fore-runner of the motor now in general use. He adopted the Otto principle, but provided for a greatly increased speed which would generate a heat intense enough to ignite the charges of gas. This was a little later still further improved by the addition of a heavy fly-wheel, and by enclosing the crank shaft in a chamber or case as is done today. This chamber, however, also contained a valve which opened inwardly and through which the explosive gas was drawn automatically by the up-stroke of the piston. When the piston began to descend the valve was closed and the charge of gas compressed within the crank chamber. Toward the end of the downward stroke a valve in the piston rod itself was mechanically opened and the compressed gas escaped into the upper part of the cylinder chamber. As the piston began to move upward again the piston valve was closed, and at the top of the stroke the charge was fired by means of an incandescent tube, heated by the great engine speed already referred to. This heat was so intense that the use of a fan was necessary to prevent the cylinder from becoming red-hot.

Former Methods of Construction.

Those who understand the working of the modern motor will readily recognize a radical change in the method of operation. The crank chamber is no longer utilized for compression, and the functions of suction, compression, etc., are performed by reverse motions of the piston. The explosive gas is also fired by an electric spark instead of a hot tube.

In 1886, Daimler fitted this engine or motor to a bicycle, suspending it vertically between the front and rear wheels. Transmission was directly from the engine to the rear wheel by means of a belt. Daimler had also fitted his machine with a primitive carbureter by which a mixture of the gasoline vapor and air was secured. This outfit, crude as it was, worked so well that Daimler continued to plan improvements, and in 1889 produced a 2-cylinder engine, the piston rods of which were coupled to a single crank. Right here Daimler's part in motorcycle construction comes to a stop. The practicability of his invention was so apparent that in the

An Old-time Motorcycle.

same year—1889—the right to make and use it was bought by Panhard & Levassor, who at once began the manufacture of automobiles—but that's another story. We are dealing with motorcycles. So far as known Daimler's interest ended when he sold his invention to the automobile men for we hear no more of him in the motorcycle field.

Another pioneer in the motorcycle field was Werner,

a Frenchman, who produced a 2-wheeled engine-driven vehicle, the distinctive feature of which was the placing of the motor immediately over the front wheel, and the transmission of the drive directly to the front wheel. In these two respects the Werner machine differed radically from the model now in general use.

Same Old Principle Retained.

Uncouth in appearance, awkward in construction, so far as the method of applying the power is concerned,

Motorcycle as Made by Werner.

the Daimler motorcycle served to illustrate a new principle in mechanics and, while there has been wonderful improvements made in details, the modern motorcycle, complete as it is, is nothing more nor less than a development of the Daimler idea. While Daimler dropped out of the limelight after disposing of his rights to Panhard & Levassor in 1889, he had set other bright men to studying the problem of gas engine propulsion for vehicles.

Experiments of Carl Benz.

Previous to 1890, antedating it by three years, Carl Benz, another German mechanic, had experimented extensively with a motor tricycle which was propelled by an oil-spirit motor, the piston being connected with a vertical crank shaft. They had queer ideas of speed in those days—and it is only twenty-four years ago—for it is recorded that Benz was greatly pleased, and the public sarcastically skeptical, when a speed of ten miles an hour was attained. Two years later, when improvements in the motor made a speed of twelve miles an hour possible, there were many prominent people who did not hesitate to say "we don't believe it." Thus Daimler and Benz were in a way contemporaneous in their experiments. The difference was that the former hitched his motor to a two-wheeled vehicle, a regular bicycle of the same form now employed in making the motorcycle, while Benz used the three-wheeled, or tricycle type.

First Motorcycle Made in America.

Lines of Development.

Strange as it may seem the line of development following the Daimler demonstrations was not in the motorcycle itself, although, as shown, the practicability of the machine was first established in this way. For some unknown reason, probably because it was thought to be more profitable, the principle was for years applied exclusively to the construction and development of automobiles and the motorcycle, for the time being, was lost sight of.

There was one notable difference in construction between the Daimler and the Benz machines, but it was in

An Odd Motorcycle.

method of arrangement only. Both employed the same general ideas as to means of obtaining propulsive power, but they applied them in a radically different manner. In the Daimler motorcycle the engine and the rest of the machinery was either in front of the rider, or beneath the frame, and between the front and rear wheels.

In the Benz vehicle the engine was placed in the rear, and drove a vertical crank shaft.

Such developments as were made immediately following the early results obtained by Daimler and Benz appear, for some years at least, to have been confined, so far as the motorcycle (or rather tricycle) is concerned, to the Benz machine. There is no actual record which would warrant this assertion, but it is a natural inference as, coming shortly after the actual Daimler motorcycle we find a machine in the tricycle form in which a part of the machinery was placed to the rear of and under the driver's seat. A considerable part of it was directly over the rear axle, and the piston or connecting rod was attached directly to the latter.

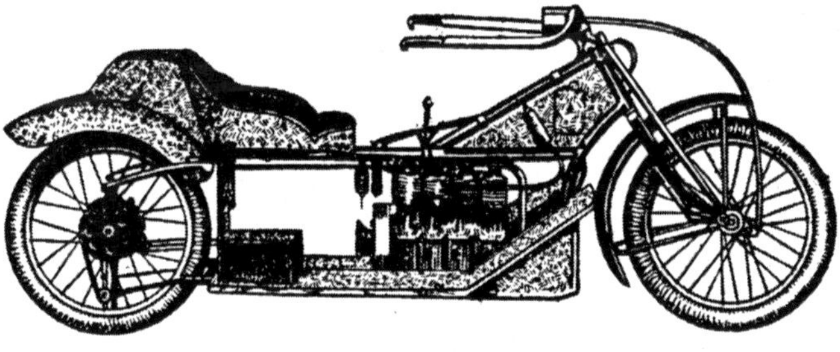

An Auto Motorcycle.

These developments, of course, set other men of inventive mind to experimenting along the same line, and one result was the production along about 1887 of a curious three-wheeled outfit.

First American Motorcycle.

The first motorcycle made in the United States (See Page 11), was constructed in 1900 by the E. R. Thomas Motor Co., which branch of the Thomas business is now owned and operated by the Greyhound Motor Works of Buffalo, N. Y. Previous to this the few motorcycles used in this country were imported. Although embracing many radical changes the Thomas machine was an outgrowth of the Daimler idea. The engine was placed against, and supported by the frame directly back of the front wheel, and the transmission, or drive, was by means of a belt running from a pulley at the bottom of the engine cylinder to the drive wheel at the rear. It will be interesting to the student of the motorcycle to note the difference between this method of construction, and that employed by Daimler and Benz, and then make a comparison with the machines of modern make.

While the improvements are little short of wonderful, and the present-day vehicle (no matter of what make it may be) an almost perfect mechanical production, the underlying principle remains the same as in 1886. In other words, the changes have been almost solely in the line of mechanical development and application. The motorcycle of 1910 does not look much like that of 1886, and yet twenty-four years ago exactly the same idea of propulsion was employed.

Great Growth in Use.

A great change has come, however, in the uses of the motorcycle, as well as in its appearance. The modern machine is something more, much more, than a mere racer, a plaything or pleasure vehicle. Within the last few years it has become in many parts of the country an essential in business and professional life. Where physicians, for instance, formerly used a horse in making their trips, many of them now ride a motorcycle. They find it quicker, cheaper, and more pleasant. In cases of emergency, distances of from thirty to forty miles may be covered as readily as six miles could be made in the days of the horse conveyance. When not in use there is no expense; no feed to buy, no care to give beyond decent shelter and an occasional cleaning. Nor is it necessary that a man should travel alone on a pleasure trip. If he wishes to take his wife or a friend along there are side-car and extra seat attachments which make it comfortable and easy.

The growth in the use of the motorcycle within the last five years has been phenomenal, and this growth is increasing yearly. Thousands are now in use, where five years ago the number could be counted in hundreds. One factory alone has arranged for an output of 10,000 machines this year (1910). It is now turning them out at this rate, and finds it difficult at times to keep up with its orders. It is apparent that the motorcycle is popular, and a permanent fixture.

In its original form the motorcycle was exclusively a one-cylinder machine. It is now made largely in the twin, or two-cylinder form, and to a certain extent four-cylinders are employed.

Motorcycle for Fire Use.

CHAPTER II.

VITAL FEATURES OF THE MOTORCYCLE.

Three vitally important systems enter into the mechanism of the modern motorcycle, each co-dependant upon the other, and each virtually useless without this co-ordination of energy. These systems are:

First—The vaporizing apparatus—the carbureter—by which gasolene is transformed into an explosive gas.

Second—The ignition apparatus—magneto or battery—by which the spark which explodes the gas is supplied.

Third—The cylinder, within which moves a piston, impelled by successive explosions of gas, by which the propulsive power is applied.

Contrary to general practice, in which the engine, or cylinder, is placed first, this is the proper order in which to name these systems for, by arranging them in this way a clearer idea as to the general outline and function of the completed machine may be had.

First we have the vaporizing apparatus because it is through this that the fuel is supplied.

Next we place the ignition, as the fuel gas is inert and of no account as a propelling force until exploded, and it is the ignition which procures this explosion.

Then the cylinder containing the piston which, energized by successive gas explosions, communicates the propelling force by which the vehicle is moved.

Each of these systems embraces various important parts, essential to their successful operation, which will be treated of in detail in succeeding chapters, but for the purpose of an understanding of the principle and operation of the motorcycle a general description of the three primary principles will be sufficient.

How the Motor Works.

As explained in the introductory chapter the modern motorcycle secures its motive power from what is known as the internal combustion engine. This is an arrangement by which a series of explosions occurring directly on the piston head within the cylinder, drive the piston down with sufficient force to carry it over the dead cen-

One Form of Carbureter.

ter, and insure its return to its original position. The lower end of this piston is connected with a crank, from which the motion is in turn conveyed by belt, chain, or shaft to the wheels. The principle is simplicity itself, but its application is the result of years of patient study and experiment. In the development of the motor this same principle of simplicity has been closely adhered to, with the result that the mechanism employed in the propulsion of the motorcycle, despite its seeming intricacy, is to-day in reality a surprisingly simple affair. Any man of ordinary intelligence may readily master the details of construction and operation, and when this is done complete understanding of what to do in case of emergency is easy.

The trouble in most instances is that the average man will not take the time to advise himself thoroughly as to the mechanism of his machine. He pays from $200 to $350 for a motorcycle and then, absolutely without

Typical Form of Magneto.

knowledge of its construction or working, beyond a bare idea of how to start and drive it, and often not even this, he will demand of it the most exacting service—service that an experienced mechanic would shrink from expecting.

It is the duty of every man who uses a motorcycle to master the principle of its construction, and become acquainted with the details of operation, not only because it will enable him to get better service out of the machine, but also because in case of stoppage or breakdown he will know what to do and thus avoid heavy repair bills.

Two Explanatory Diagrams.

The diagrams given illustrate the principle of motorcycle construction and operation. Individual makers have their own ideas as to methods of generating and applying power, form of frame, etc., but the underlying principle in all machines is the same. Gasolene is taken from a storage tank, put through a carbureter to vaporize it by securing a mixture with oxygen, then introduced

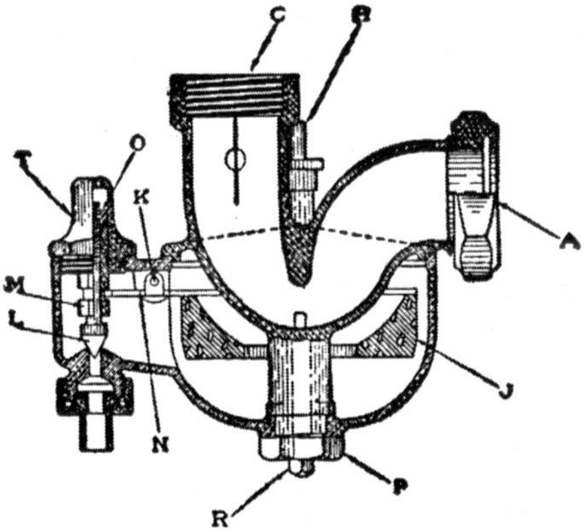

The Holley Carbureter.

into the cylinder where it is exploded by a battery or magneto spark, these explosions in turn moving a piston which revolves a crank propelling the wheels. This, in brief, is all there is to the motorcycle. Comparison of the diagrams with the accompanying key will make its operation plain. See diagrams on pages 21 and 23.

Value In Knowing How.

Here we have a compact, easily understood description of the practical workings of the machine from cause to effect so far as primary principles are concerned, and we must master these first before we can go further. It is just like a child learning his A B C's before he begins to group them into words and then into sentences. In succeeding chapters we will learn how energy is produced, how it is applied, and what the result is. This done we will have mastered the mechanical study of the motorcycle. From then on it will be merely a matter of experience and observation, but this latter

Excelsior Air-Cooled Cylinder.

field is by no means a small one. No matter how perfect a man may be in the knowledge of his machine new things are continually arising to test the ingenuity of his resources. It is the man who has mastered the mechanical details that is best equipped to successfully meet these problems as they arise. More than this the mechanical faculty, once fully aroused and developed, impels a man to research on his own account, often resulting in the invention of devices of great value.

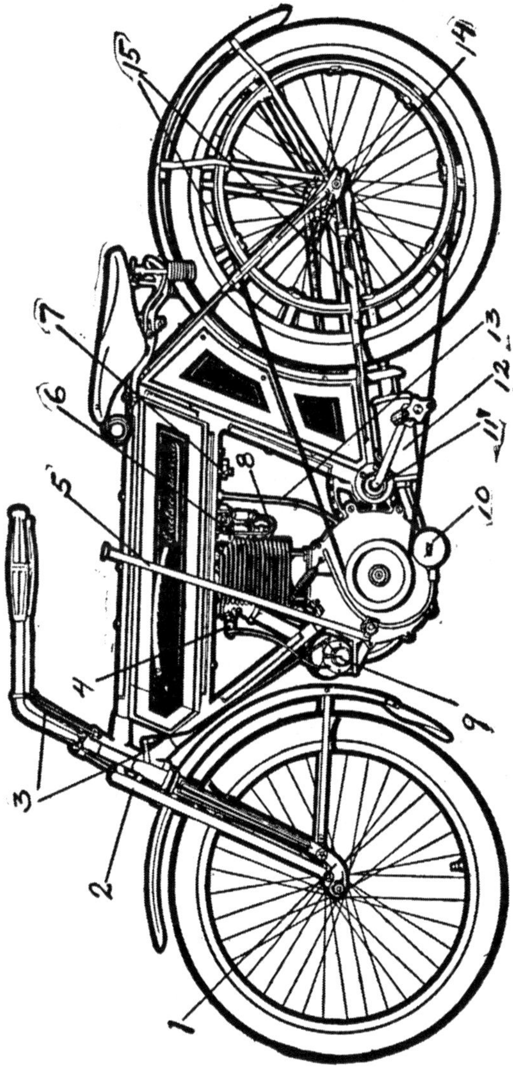

Parts of Excelsior Single Cylinder Motorcycle.

Location of Motorcycle Parts.

It is impossible to give any one description of the assembling—the putting together of the various parts—that will answer for all machines. Each manufacturer has his own ideas on this point and, as a consequence, it is rarely that two machines of different make are put together exactly alike. Owing to this it is impracticable to say just where the carbureter, the magneto, or any other vital part is located. In one machine, for instance, the carbureter may be directly over, or alongside of the cylinder, while in another it will be found well back of the cylinder. In some machines the gasolene is carried in a tank suspended from the top of the frame, in another it may be attached to the rear part, while in at least two instances the frame itself is made to serve as a tank, being composed of large hollow tubing holding about a 2-gallon supply. The only thing to do in the way of indicating the various parts so the novice may identify them is to take some one make and point them out. This is done with the one and two cylinder Excelsiors, cuts of which are printed with this chapter. By reference to the following numbered "key," and comparison with the same number on the cuts, the location of any important part may be ascertained. There are some parts, such as the cylinder, the wheels, etc., which it is unneccessary to number as they show for themselves.

One Cylinder Machine.

1—Spring fork rocker arm.
2—Auxilary fork.
3—Spark and throttle controlling mechanism.
4—Spark plug.
5—Belt tension lever.
6—Carbureter.
7—Gasolene draw-off cock.

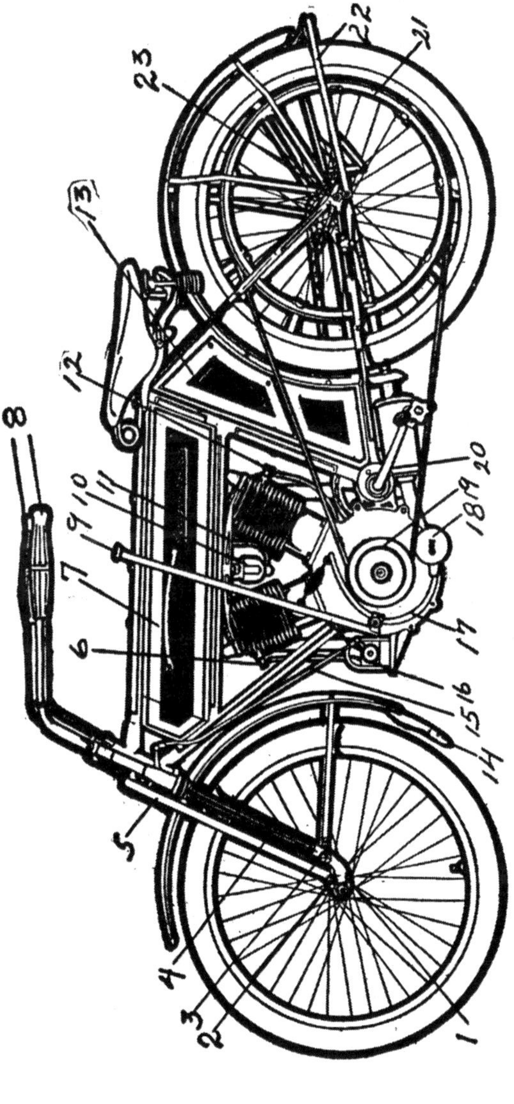

How the Excelsior 2-Cylinder Machine Is Put Together.

8—Gasolene pipe to carbureter.

9—Magneto.

10—Idler pulley.

11—Eccentric crank hanger for adjusting drive chain.

12—Muffler.

13—Oil pipe from oil reservoir to crank case.

14—Rear wheel adjustment for adjusting belt tension.

15—Offsets in rear forks to allow belt to pass.

Two Cylinder Machine.

1—Rocker arm for spring fork.

2—Taper studs at lower end of auxiliary fork.

3—Taper studs joining rocker arm to main fork.

4—Fixed or main fork.

5—Movable or auxiliary fork.

6—Spark plug forward cylinder.

7—Gasolene and oil tank.

8—Rotable hand grips for steering and motor speed control.

9—Idler lever for adjusting belt tension.

10—Carbureter.

11—Truss for securing engine heads to frame.

12—Spark plug, rear cylinder.

13—Tool and luggage box; also used for batteries and coil on battery driven models.

14—Splasher or boot on front mud guard.

15—Spark controlling mechanism.

16—Magneto.

17—Belt tension regulating mechanism.

18—Belt tension pulley.

19—Motor pulley.

20—Eccentric hanger for adjusting belt tension.

21—Belt sheave for driving pulley on rear wheel.

22—Stand.

23—Rear wheel adjustment for regulating belt tension.

CHAPTER III.

MOTIVE POWER—PRODUCTION AND APPLICATION.

As briefly outlined in the preceding chapter, the force, or motive power which propels the motorcycle is obtained primarily by the successive explosions of charges of gasolene vapor inside the engine cylinder, and directly on top of the piston head. This piston head is connected with a crank shaft which in turn is connected by chain, belt or shaft with the rear wheel. As the piston is forced downward by the explosion of the gasolene vapor the connecting rod revolves the crank shaft, and this motion which, as the explosions grow more frequent and regular becomes continuous, is communicated to the rear wheel by a carefully studied-out system which insures an increase in the velocity and produces a high speed. The reason for applying the motive power to the rear wheel is that the front one must be left free of rigid or direct connection with the motor so as to permit of its use for steering purposes.

Gasolene the Exclusive Fuel.

In the production of this motive power gasolene is almost exclusively used as a fuel solely because of its convenience and economy. There are other fuels that would produce the same results, but none so cheaply or effectively. Extensive experiments, and in some instances, practical operations have been conducted with alcohol as a fuel, but thus far the use of alcohol has not been seriously considered for two reasons. One of these is that even in its denatured form, it is more expensive than gasolene, costing about double on the basis of results obtained. As the tendency is continually toward a reduction in the expense of production, however, this feature of added cost may be in time overcome. A much more troublesome feature is to be found in the fact that alcohol will not vaporize sufficiently without the application of heat, and it is therefore necessary to use gasolene in the starting of the motor, until heat enough is generated to insure continuous vaporization. This makes it necessary to carry two fuels, to use one first, and then shut it off when conditions are right for the introduction of the second.

Despite these drawbacks it is probable that human ingenuity will in time overcome them, and that alcohol may become the motorcycle fuel of the future. For the present, however, we must deal exclusively with gasolene.

Kerosene and Benzene as Fuels.

Considered solely from its heating, or power-producing results, kerosene is a cheaper fuel than gasolene, but here again we run against the objection of deficient vaporization, and in the successful operation of an internal combustion engine we must have a ready and thorough vaporization of the fuel. Kerosene, like alcohol, requires the application of heat to produce this. On an intensely hot day it will work well, but at normal temperature the vaporization is very slight, and at low temperatures exceedingly difficult to procure. Besides this kerosene contains more carbon than gasolene and as a consequence fouls the cylinder more.

English experts are experimenting with benzene, or more properly speaking, benzol. If it could be obtained in sufficient quantities and at a fair price, say about the same as that asked for gasolene, it would doubtless become a popular fuel. It has a greater heating value, and consequently more power than gasolene; it is cleanly, and does not lose its strength so quickly by exposure to

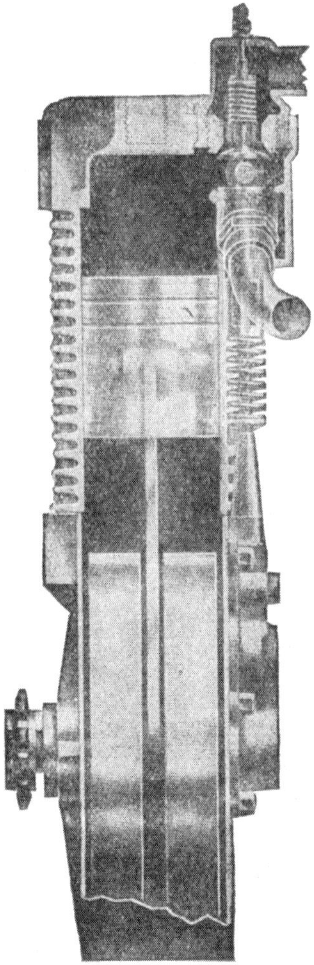

Action of Connecting Rod.

the atmosphere. While it is not so volatile as gasolene, and does not vaporize so rapidly, it can be made to do the work, and a motor may be started with it without difficulty. The production of benzene is, however, limited, and despite the fact that gallon for gallon it will produce more power than gasolene, it is at the present time too expensive. The commonly accepted idea is that benzene is a product of the distillation of petroleum, but a comparativly small quantity is obtained in this manner. It was originally derived from coal tar, but the modern process is to get it from coal, about three gallons being secured from every ton of coal used in the production of gas or the manufacture of coke. In cooking the coal, either to obtain gas or coke, the gases are collected and condensed, one of the products being benzene.

Chemists Seek Other Fuels.

There is a good reason why chemists are continually at work seeking to produce a substitute for gasolene. At the present, and ever-increasing rate of consumption, this fuel is bound in time to become scarce and advance in price. It is one of the products of petroleum, about 8 to 10 per cent of the distillation of the petroleum being gasolene. The other products are naptha, 8 to 10 per cent; kerosene and light lubricating oils 70 to 75 per cent; heavy residium 5 to 10 per cent. The production of gasolene must, therefore, as a commercial economy, be restricted to a point where there will be no surplus of the other products which form the great percentage of the distillation of petroleum. As a commercial proposition it is impossible to distill petroleum with the main purpose of obtaining gasolene unless all the naptha, kerosene, lubricating oils, residium, etc., can be marketed promptly at a satisfactory price. Chemists understand that there is a limit to the demand for these latter articles which in turn tends to limit the production of gasolene. When this is reached the price will advance, and other fuels will come into use.

Proper Mixtures To Use.

In generating motive power it will be well to remember that there is such a thing as overdoing it. Because, as a mathematical proposition, a gallon of gasolene will produce four times as much power as a quart it does not follow that when four times the power produced by the

GASOLINE VAPOR AND AIR.	TIME IN SEC. BETWEEN IG. AND H. PRES.+	EXPLOSIVE FORCE IN POUNDS PER SQ. IN.			TEMPERATURE OF COMBUSTION IN F. DEGREES.	
		COMPRESSION IN POUNDS PER SQ. IN.				
		45	60	80	ACTUAL	THEORY
1 TO 13	0.28	156	208	260	1857	3542
1 TO 11	0.18	183	244	305	2196	4010
1 TO 9	0.13	234	312	390	2803	4806
1 TO 7	0.07	261	348	435	3119	6001
1 TO 5	0.06	270	360	450	3226	6884
1 TO 4	0.07	240	320	400	2965	5517
+ AT ATMOSPHERIC PRESSURE.						

Explosive Force of Gasolene and Air.

quart is required it is necessary to use the gallon. One of the abuses to which the motorcycle is subjected is the tendency on the part of inexperienced riders to use too much gasolene. It is indisputable that a "rich" mixture will ensure more explosive power than a "lean" one, but it does not follow that it is always wise to employ the maximum of "rich" mixture when an unusual amount of power is called for.

By reference to the accompanying table on the "Explosive Force of Gasolene and Air," it will be seen that a mixture of 1 part of gasolene with 13 parts of air, exploded under a compression of 45 pounds to the square inch will produce an expansive force of 156 pounds per square inch. At first thought it might be taken for granted that a mixture three times as rich, that is, one containing only 4 parts of air to 1 of gasolene would produce a force three times as great, or 468 pounds to the square inch, but it does not figure out that way. The real result of the 4 to 1 mixture is only 240 pounds of force. We are thus securing a little more than one half more force by using more than three times as much force-producing explosive.

Nor does the rapidity of explosion produce the increase in power which is generally supposed, at least not the proportionate increase. In the example quoted the mixture of 1 to 13 exploded at intervals of 0.28 second produced 156 pounds of force, while the 4 to 1 mixture which produced only 240 pounds was exploded every 0.07 second

The lesson of this is that there is a limit to the economical use of gasolene in producing power and speed. The fact is that excessively rich mixtures are of value mainly in starting or in running slowly, and then only because there is less trouble in igniting them. It is sometimes difficult to secure the ignition of a lean mixture, especially before the machine is under headway, or is moving at a low speed. In a condition of this kind the experienced operator will increase the flow of gasolene and then, when the motor is working well, reduce it to the minimum that will secure the desired results. Most men of this class prefer the leanest possible mixture that will keep the machine moving regularly and under the required headway. They realize that it is a waste of effort and money to burn a gallon of gasolene if a quart will answer the purpose.

This, of course, does not apply to racing contests where extreme speed is the sole requirement, and expense of operation is not to be considered. In contests of this nature while using three times as much gasolene may give a little more than one-third more power this additional energy, small as it is, may be the deciding factor.

Application of the Power.

With the gasolene vapor tranformed into power it becomes the work of the motor to turn this power into

Interior of Crank Case, Showing Fly-wheel.

active energy and apply the energy to the wheels—in other words to "make them go round." The motor, as explained in detail in Chapter VI under the heading "The Motorcycle Engine," consists of an airtight cylinder within which moves up and down a snugly fitting piston head. From this piston hangs a connecting rod, the lower end of which is attached to a crank on the shaft of which is a geared wheel. Within the crank case, and fitted into the main gear are other geared wheels. As the crank is revolved by the up and down motion of the connecting rod these gears in turn give automatic motion to various valves and other parts. One end of the crank shaft projects through the crank case and on this is fitted a small pulley or wheel if the drive or transmission is to be by belt or chain. If a shaft drive is used the transmission from the crank is by gears.

Up to the point of delivering the energized power at the crank there is little or no real difference in the principle employed in the various makes of motorcycles. From here to its connection with the hub of the rear wheel where the travelling motion of the machine begins the methods in use are many and varied, each having ardent supporters who will contend that it is the one and only method by which satisfactory results may be had. There are flat belt and V belt, chain and shaft drive enthusiasts, each positive that the system he advocates is the best, but from the fact that all systems continue to be employed after years of experiment, and are growing in demand every year as the sales of motorcycles increase, it is evident that all have merit which finds favor among the men who ride the machines.

CHAPTER IV.

WHAT THE CARBURETER DOES.

Before the fuel power of gasolene can be turned into a propulsive force the gasolene must be vaporized—mixed with air. This carburation, or mixing with air produces

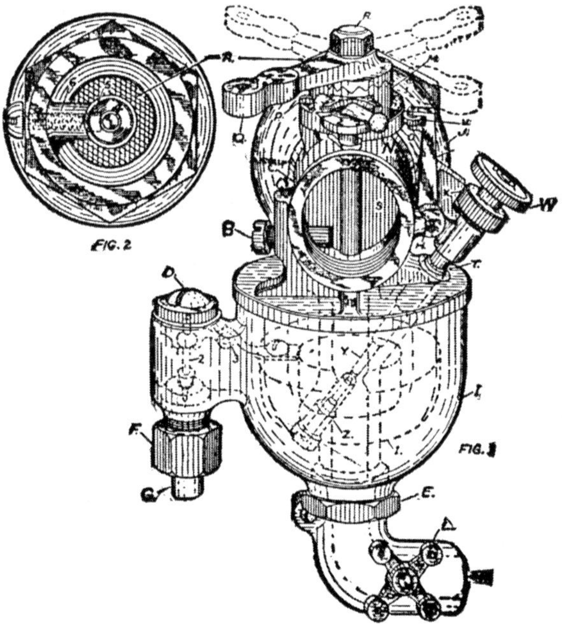

Schebler Motorcycle Carbureter.

an explosive vapor or gas. This is what the carbureter does, and the name by which the device is known is self-explanatory. In doing this it performs a duty similar to that of the heart in the human body, supplying, as it were, the life blood of the mechanism.

While there are several forms of carbureters the main

principle of action is the same in all. The supply of gasolene is carried in a tank placed well above the carbureter so that gravity will ensure a steady feed, the exact amount being regulated by a cock which is manipulated by the rider at will. From this tank the gasolene flows through a feed pipe into the lower part of the carbureter and then through a spray nozzle which discharges it in the form of a fine spray or vapor into an upper chamber. In the upper part of this latter chamber is an air intake valve, insuring a constant supply of fresh air. The gasolene vapor and the air mix, forming an explosive gas which produces the propulsive power.

A feed pipe connects the carbureter with the top of the motor cylinder at one side. This cylinder, of course, is air and gas tight. As the piston begins its downward stroke a suction is created by which the gas-and-air mixture is drawn into the top of the cylinder through an automatic check valve. Here the duty of the carbureter ends, and the work of propelling the vehicle devolves upon the ignition service and the cylinder action which are described in other chapters.

The Forms In General Use.

Thus far only one or two firms of carbureter makers are supplying this vital part of the motorcycle equipment, but it is probable others will soon enter the field, as the motorcycle industry is becoming too important to be overlooked. In some instances the motorcycle manufacturers themselves have produced satisfactory carbureters, but for the purpose of illustration the Schebler is used, not so much because it possesses any marked advantages in principle over other makes, as from the fact that, being produced by a firm which turns it out as a distinct specialty, it has been found in many ways one of the most reliable for motorcycle use.

The Schebler, like all modern carbureters, is of the float-feed type, this design having virtually superseded the old-time surface mixer. Being in almost universal use, the motorcycle user should make a close study of the accompanying diagram and explanatory key. This will give him a thorough understanding of the working of the mechanism and enable him to intelligently regulate the operation of the various parts should it become

MODEL D

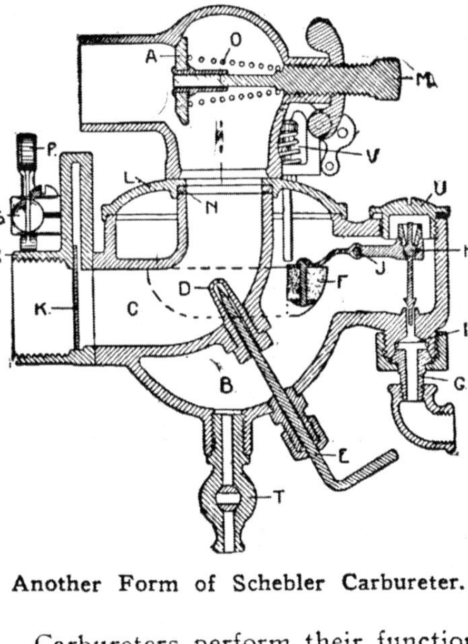

Another Form of Schebler Carbureter.

necessary. Carbureters perform their function mechanically, and as a rule do it well, but there are occasions when, from some foreign obstruction, they will fail to work satisfactorily. The man who has a thorough knowledge of the mechanism and operation of this part of his motorcycle is naturally in better position to overcome the trouble and keep his machine going than the one who does not understand it. More than this, he not only has an uninterrupted use of his machine, but he saves a lot of money by avoiding useless repair bills.

How to Understand the Carbureter.

In order to get a clear idea of the interior workings of the carbureter the various parts will be described in their order of operation, instead of by alphabetical sequence. Thus we start with the introduction of the gasolene into the carbureter and follow it through the various stages until the vapor is discharged into the cylinder ready to be exploded and do its work in propelling the machine. This forces an arbitrary use of key letters, G, for instance, being used first because it designates the introductory feature of the operation. The person who will study the following keyed description closely, at the same time making reference to the diagram of the interior mechanism, will readily get an intelligent, comprehensive knowledge of carbureter operation and this is essential, not only to ensure satisfactory results, but as a matter of personal satisfaction. Reference is made by letter only to such parts as are of vital importance in the working of the device. There are several parts lettered on the diagram which it is not necessary to consider at this time.

G.—Connection with feed pipe from gasolene tank. Gasolene is fed to carbureter through this.

F.—Float which regulates supply of gasolene. This is made of some light material, usually cork, and rises and falls with the level of the gasolene supply. When the float goes down the valve H is opened, and the gasolene comes in; when the float is up the valve is closed, and the supply is cut off.

B.—Float chamber in which gasolene is held previous to being vaporized.

D.—Spraying nozzle. No gasolene can get into the mixing or vaporizing chamber C, except through this nozzle. The head of this nozzle is full of fine holes through which the gasolene, by gravity pressure, is forced into the mixing chamber in the form of a spray.

E.—Handle of needle valve by which the flow of gasolene through the spraying nozzle "D" is regulated. When a "rich" mixture is required the flow of gasolene is increased by turning this handle to the right; when it is desired to make the mixture "lean," the flow of gasolene is reduced by turning the handle in the opposite direction.

C.—Mixing chamber in which the vaporization process takes place.

A.—Air intake valve. This admits air in quantities as desired, the flow being governed by the thumb screw M. The air passes down into the chamber C where it mixes with the gasolene vapor, forming the motive power gas.

K.—Throttle valve. This is operated by the lever P, which is worked automatically by the cylinder piston movement, and controls the flow of gas through the pipe R into the motor cylinder.

T.—Drain cock through which the carbureter may be emptied and all dirt, sediment and water removed. This should be done occasionally as water is liable to get into the gasolene occasionally, and frequently there is obstruction from dirt particles no matter how carefully the gasolene may be strained.

This is not the model generally used on motorcycles, but the difference is mainly in size and form, the principle of operation being the same in each. Model D was selected for illustration merely because the working plan is more distinct. The one used on motorcycles is known as Model H.

Regulating the Gas Mixture.

It is a well-understood fact that the more work an engine is called upon to do the more fuel it must have. It is also equally well understood that the better the grade of fuel the easier this work will be performed. Thus coal of high grade will make more steam, and as a consequence develop more power, than that of inferior grade. Some motorcycle experts say this does not apply to gas motors and that the matter of increased power and speed is to be obtained solely by increasing the volume of gas passing into the cylinder regardless of its quality. To this theory the author can not subscribe in full. While volume is a necessity, so is quality as, without this latter the explosions, while rapid enough, will be lacking in power. Take, by way of illustration, the charge of powder in a shot gun. Three drams will propel shot of a given weight at a certain velocity; an increase in the powder charge to four drams will ensure a higher velocity, and consequently a greater impact, or striking force. But, of course, there's a limit to the amount of powder which can be used safely and economically, and this also applies to the gasolene vapor.

It is through the carbureter that an increase of speed and power is supplied when desired. Or rather, to speak more correctly, it is through the carbureter the vital force which produces this additional speed or power is obtained. Motorcycles not called upon to carry excessive loads or climb sharp hills, can get along very well on a minimum amount of fuel force, when once under headway. As the work which the motor is called upon to do is increased the volume of gas should be increased, and sometimes enriched by the addition of more gasolene. This is accomplished by means of the thumb screw M, and the valve lever E. For instance, if the rider is called upon for an extra exertion in the way of climbing a hill, carrying an unusual load, or extra speed, he turns the lever E so as to increase the flow of gasolene, thus making a "rich" mixture by injecting a larger proportion of the gasolene vapor. If, on the contrary, the conditions are such that the minimum of mechanical or motor effort is sufficient he reduces the gasolene supply by means of the lever E, and at the same time increases the air supply by turning the thumb screw M.

It must be kept in mind, however, that there is such

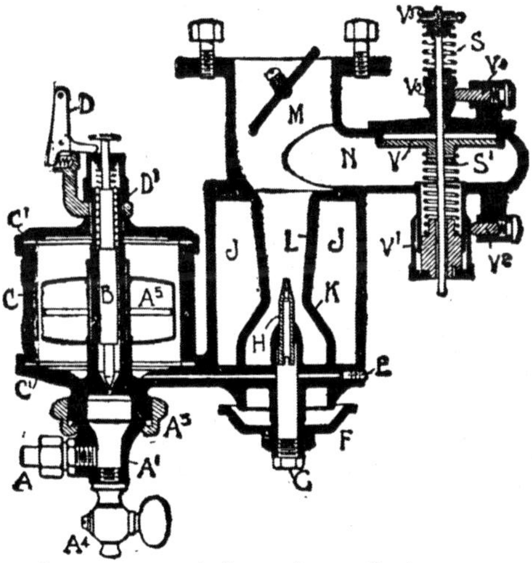

One Type of Stromberg Carbureter.

a thing as making the mixture too rich. This does not increase the power or speed of the machine, but on the other hand tends to retard progress by causing sooty spark plugs and foul cylinders. An excessive amount of gasolene clogs the cylinders with carbon, causing what is called "pre-ignition." It also coats the spark plug with soot and makes ignition uncertain and difficult. There is a maximum degree of "richness" which a machine may utilize economically, and this can be best determined by watching the exhaust. If the escaping smoke is densely black it is a sign that too much gasolene is being used. The same is true if a red flame issues from the exhaust.

The Necessity of Air.

While air in itself has no explosive or force-giving power unless highly compressed, its mixture with gasolene vapor is absolutely essential to insure the full combustion and explosive power of the latter. Thus, we know that ordinary illuminating gas will not burn in a vacuum; it requires a certain percentage of oxygen to obtain the best results in the way of combustion and resultant illumination. This same theory holds good in the combustion of the gasolene vapor from which the motive power of the motorcycle is obtained. Just what percentage of air is necessary is problematical, but in a general way it may be given as 160 cubic feet of air (at 60 F. and atmospheric pressure) to insure the complete combustion of one pound of gasolene. There is great variation in the quality of this fluid, depending upon the nature of the petroleum from which it is obtained, and the process of distillation employed.

How to Test Gasolene.

The quality of gasolene is tested by specific gravity with an instrument called a hydrometer. This is a simple, cheap apparatus which every motorcycle owner should possess. By its use he can protect himself against fraud, especially in the admixture of water, a practice which is not uncommon among unscrupulous dealers. The hydrometer is a glass instrument on the face of which is a scale. When the hydrometer is placed in a tank or pail of gasolene the specific gravity of the latter is accurately recorded. This specific gravity is computed on the initial basis of the weight of water. Gasolene, for instance, that is of .74° specific gravity, weighs 74.100 what the same quantity of water would. The most preferable grade of gasolene for motorcycle use is that not lower than .73° to .76° specific gravity.

Gasolene is much lighter than water and, when not agitated, the latter will settle to the bottom. It requires only a few drops of water in a gallon of gasolene to interfere with the proper action of the latter. Hence the necessity for keeping a close watch and making frequent tests.

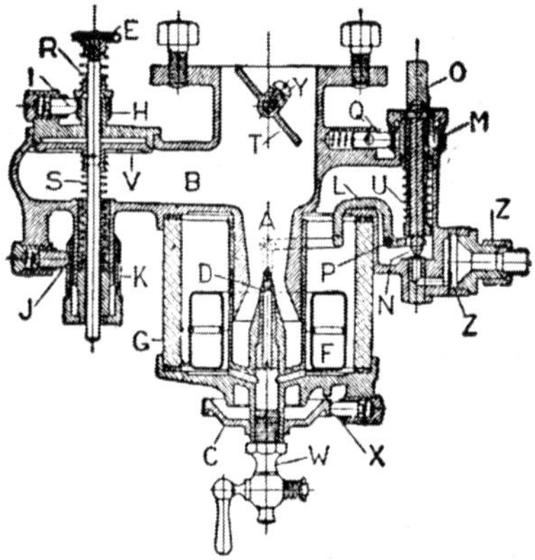

Type B, Stromberg Carbureter.

A simple means of detecting water in gasolene is to pour a little in the palm of your hand, or on a good-sized coin like a silver dollar. Let the gasolene evaporate. If it contains water, drops of the latter will remain.

There is only one way to separate water from gasolene and that is to strain it through chamois.

Use of Throttle Valve.

Increased power and speed are also to be obtained by opening the throttle K wider, and allowing a freer passage of the gas into the motor cylinder. It would do little good to enrich the mixture if provision were not made for feeding it to the cylinder in enlarged quantities. This regulation by the throttle is necessary if full advantage is to be taken of the additional gasolene used. To make a richer mixture and not provide for increased consumption in the cylinder would put the motorcyclist in the position of an engineer who was overcrowding his boiler with steam, and not bringing its expansive force to bear upon the piston head. More power and speed may be obtained to a limited extent by merely opening the throttle without changing the mixture, but the most satisfactory results will follow the combination within reasonable limits, of a richer mixture, and a more liberal feeding of the gas to the cylinder. In all modern motorcycles there is a throttle control connected with one of the handle bars so that the rider may open or close, or otherwise regulate the throttle by simply pressing a lever alongside the handle grip, or turning the handle grip itself.

Leave the Carbureter Alone.

Considering the important part it holds in motorcycle mechanism the carbureter is surprisingly insignificant in size. The average is about 3 by 5 inches, with a weight of from 10 to 12 ounces. Despite this it is not of delicate construction, being stoutly and strongly made, and performing its duty with clocklike regularity and dependency. It is seldom that serious troubles originate in the carbureter through any defect in mechanical operation. Not understanding this there is a tendency in case of stoppage to look for the cause in the carbureter itself. This is inexcusable, as it generally results in more aggravating and expensive delay than would have come from the original cause if it had been properly located and treated. About the only serious trouble in the carbureter itself may be traced almost invariably to dirt in the gasolene. To insure constant operation of the motor the supply of gasolene must be continuous. The moment this supply is cut off the motor naturally stops running and there is a so-called "breakdown." Gasolene should be carefully strained when put into the tank before starting on a trip, and at every refilling, but even when this is done more or less dirt will get in. The holes in the spray nozzle, for instance, are very small,

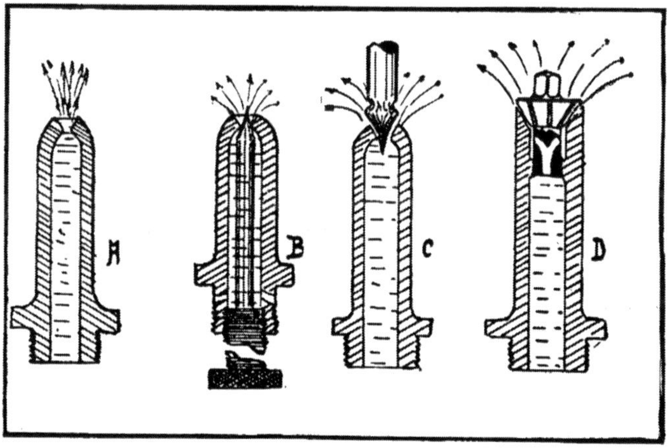

Various Forms of Spray Nozzles.

and minute particles of dirt will clog them. One little piece, no larger than the head of a pin, has great trouble-making power. When this stoppage is obstinate there is only one thing to do—drain off the gasolene, and clean out the carbureter thoroughly. The necessity for doing this, however, will not be so frequent as the amateur may imagine. Once, or at most twice, a season should be enough provided due care is exercised in the selection and straining of the gasolene.

In Care and Repair of the Motorcycle, published by the Bicycling World Co., the following amusing incident, showing how a little thing may cause a lot of trouble, is given:

"One case is known in which for more than three months a machine cut all manner of erratic capers, the cause of which baffled a dozen experts and nearly led to a law suit, and which ultimately was found to be due to a dead fly which had lodged in the feed pipe of the carbureter."

There's one safe rule to follow: Look for trouble in every other part of the machine, particularly the electric system, before you investigate the carbureter. There is less real trouble here than most people think.

Another thing—don't attempt to take the carbureter apart until you understand its mechanism thoroughly, and are sure you can replace the various parts properly. Don't let the roadside tinker, the man who hangs out a sign "Motorcycles Repaired," dissect your machine unless you know that he is capable. Very few of these men really understand its mechanism and make stupid blunders. In one instance one of these tinkers replaced a float upside down, and of course the engine only fired intermittently. Considerable money was expended in vain efforts to locate the trouble, but it was not until the machine was sent back to the maker, and got into the hands of an expert, that the real trouble was discovered.

Connecting and Adjusting Carbureter.

Although the carbureter will be properly attached to the machine when it comes from the factory it is well to understand how to make the connection yourself in case of emergency. This is explained in the case of the Schebler carbureter as follows:

Connect carbureter to intake pipe so that it sets about

six inches below bottom of gasolene tank, that the bowl may be filled by gravity. For best results, the carbureter should be as close to cylinder as possible, and in case of multiple cylinders, equidistant from each one. Connect a pipe or tube from gasolene tank to union "G." Pipe to be brass or copper and not less than 1/8 inch hole. (N. B. Do not use rubber tube, as float chamber will not fill properly and rubber makes a sediment in gasolene.) Be sure before connecting, that the gasolene tank has no scrapings, filings or other particles of dirt on the bottom, and that the tube or pipe is thoroughly free from dirt or corrosion.

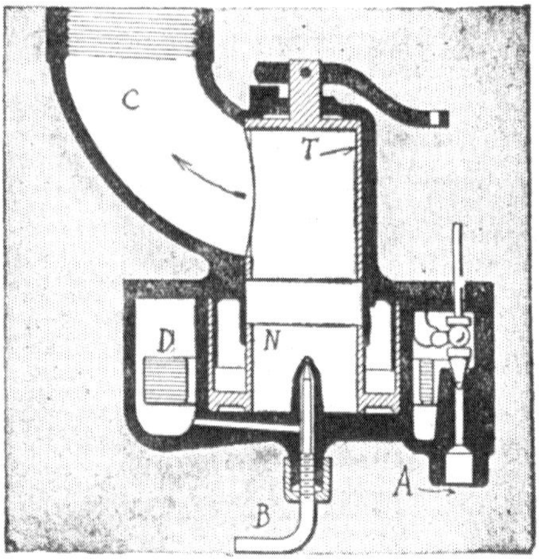

Section of Duplex Carbureter.

After connections are properly made, turn on the gasolene and wait a few moments for float chamber to fill. The needle valve should be closed and then opened about five-eighths to three-fourths of a turn. Retard spark (see explanation under "Ignition"), open throttle K. about one-fourth, so as to equal the size of the fixed air opening. Start motor and adjust needle valve E until motor runs smoothly without back-firing or missing. Now open the throttle wide, keeping spark retarded, and by turning air valve adjusting screw M, increase or weaken tension on air valve spring O, until proper mixture is obtained. If the mixture is too "rich," the tension of the spring should be weakened, allowing more air to enter the carbureter, and if the mixture is too weak, tighten the tension.

Remember, that to insure successful operation the

Another Form of Carbureter.

float chamber C must be full. This can be accomplished by pushing down on the flush pin V, which depresses the float, thus opening the float valve H, and allowing the gasolene to enter.

A Few Carbureter Troubles.

When starting your machine watch the muffler cut-out. or direct exhaust. If red flame and black smoke come out you are using too much gasolene. Your mixture is too rich. Cut off some of the gasolene.

Occasionally, but not often, the float valve will leak, the air valve action become weak or the fuel adjustment clogged. Any of these will manifest themselves in the production of a rich mixture by allowing of the introduction of too much gasolene, and when the rider is certain that his needle valve is set right it is easy to trace the cause of the superfluous flow to one of the

Still Another Form of Carbureter.

troubles named. When the source is located any man who has studied the working of the carbureter up to this point can readily overcome the trouble.

A leaky intake pipe, clogged spray nozzle, air vent to float chamber stopped, dirt or water in float chamber, or fuel adjustment loosened will also manifest themselves in a weak mixture due to the absence of the necessary amount of gasolene, and the presence of too much air. If the gasolene tank contains an adequate supply of fuel, and the needle valve is set properly then you may look for the cause of the weak mixture in one of the troubles named.

Aside from these there are few, very few, troubles which may be traced to the carbureter.

If a yellow flame comes from the exhaust the mixture is not rich enough, but this is not properly a carbureter trouble, although it is to be remedied through the action of this device. Open the needle valve slightly to admit more gasolene.

Sometimes when starting the engine on a cold morning there will be a little trouble from what is known as "back-firing" (this is explained in the chapter on engines). This is caused by the metal of the various parts being cold, and hence contracted. When this occurs do not attempt to readjust the carbureter. Wait until the engine has become warmed up by the explosions of the gasolene vapor when the back-firing will cease, and the operation become normal. Some riders hasten the starting of their motors on cold mornings by injecting a few drops of gasolene through the petcock on the cylinder, provided there is one. In the absence of a petcock dip a rag in hot water and wrap it around the carbureter and inlet pipe. This will hasten the vaporization of the gasolene, and save a lot of hard pedalling in getting under way.

CHAPTER V.

IGNITION—BATTERY AND MAGNETO.

Having vaporized the gasolene, and introduced the vapor into the cylinder, the next move is convert the vapor into force-producing power by exploding it. Bear in mind the important fact that it is these explosions, rapidly recurring which cause the piston to move and thus communicate motion to the wheels of the machine. It is by ignition that these explosions are secured.

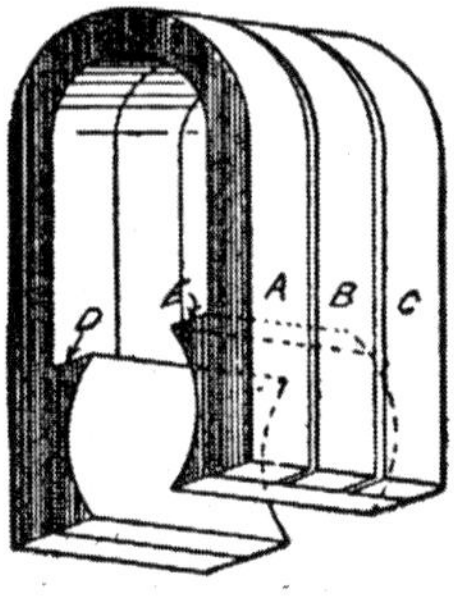

Magnets in a Magneto.

And in this instance the word is used in its actual, literal sense—a kindling or setting on fire. While the gasolene vapor is the fuel from which the motive power by which the machine is driven is obtained, this vapor can exert no influence in the direction of propulsion until combustion, or explosion occurs. Hence the ignition system becomes a highly important factor in the successful operation of the motorcycle. We must have absolute certainty in the ignition, a chronomoter-like regularity in the succession of sparks, or the engine will not work properly.

Aside from the matter of convenience it makes no particular difference by what method this ignition is secured. Anything that will bring a sufficient amount of

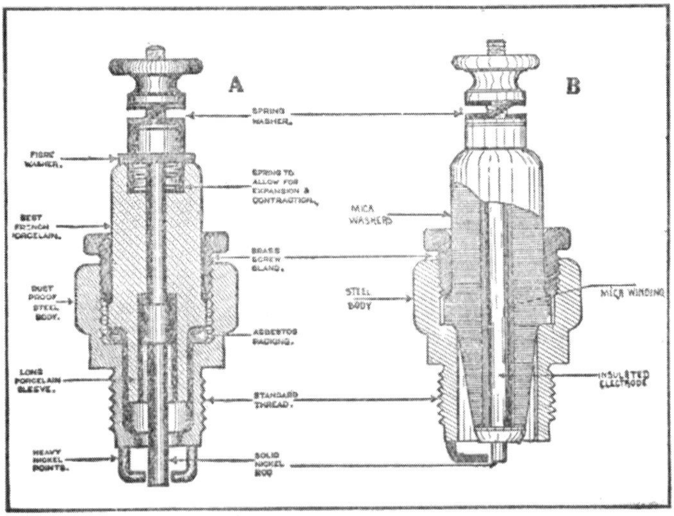

Porcelain and Mica Spark Plugs.

heat or flame to bear upon the vapor will do the work. In the early days of the gas motor several widely differing forms of obtaining ignition were utilized. In one instance it was an actual flame; in another a heated tube. Nowadays the electric spark is exclusively employed, solely as a matter of convenience. It is the simplest, surest way of obtaining the desired ignition.

There are two methods of producing this electric-spark ignition—the battery and the magneto, both of which are described in detail further on. Both of these systems, while radically different in their nature, produce exactly the same result—a small electric spark in the cylinder which explodes the gasolene vapor. Each system has its advantages and its drawbacks, and each has its advocates and its opponents. Those who use the battery system will insist that the magneto is too complicated and consequently too liable to get out of order.

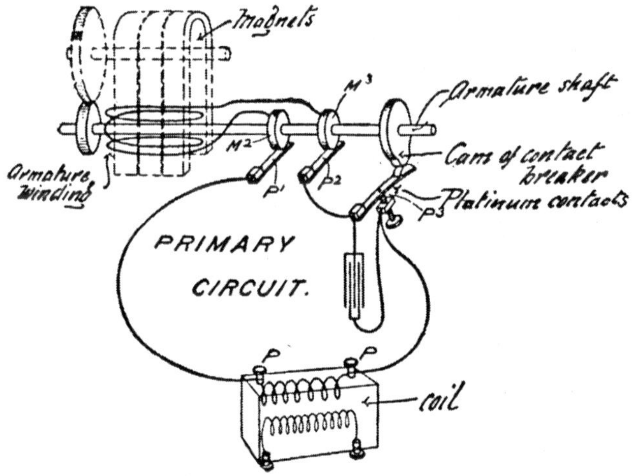

Primary Circuit of Magneto System.

The magneto advocates, on the other hand point to the fact that batteries require the use of a spark coil, have to be replaced occasionally, and that this and the necessity of frequent inspection makes them a nuisance. In this connection it should be remembered that the battery has the advantage of being the older system, and is naturally more highly developed. Improvements are being made in the magneto right along, however, and it is rapidly approaching a stage of perfection.

How the Magneto Works.

Magnetos as used on the motorcycle are of the horseshoe shape. All of us can readily recall the toy magnet of childhood by which needles and nails, and similar articles of iron and steel, could be lifted. The magneto is simply the toy magnet enlarged and operated mechanically so as to produce an electric current. To insure greater efficiency the compound form of magnet is used.

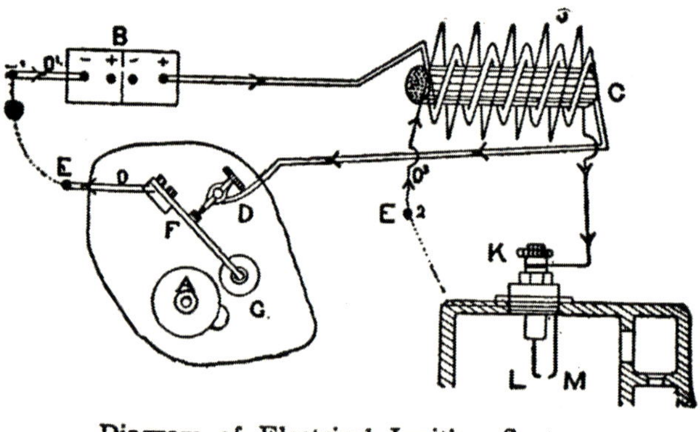

Diagram of Electrical Ignition System.

The key letters designating the various parts are the same in this as in the other cuts. The additions are K connection with spark plug; L M wires in jumping the gap between which the spark is produced.

This consists, as shown on page 54, of a number of strips of steel bent into the shape of a horseshoe, and placed against one another, either lapped over one another on top, or arranged so as to form an arch. While all of this surface has more or less magnetic energy the attraction is always strongest at the poles or ends.

Mechanical generators of electrical energy used as a source of current for ignition are of two types, although the basic principles upon which their construction and operation depend are the same. They are known as dynamos and magnetos, and can be readily distinguished from each other by distinct peculiarities of construction. They produce exactly the same results, though operating in a radically different manner. Most manufacturers prefer the magneto to the dynamo because of its

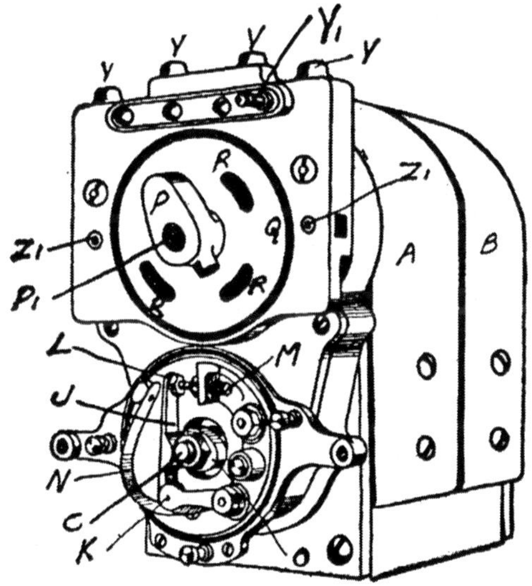

Front View of Eismann Magneto.

A B magnets; C armature shaft; D E gear wheels which act as high tension distributors; J K L M various parts of the contact breaker; P rotating arm which receives the high tension current from the coil; R R R R segments which when pressed against the disc L by the arm P "make" the current; Y Y Y Y terminals from which wires conduct the current to the cylinders.

more compact form and lighter weight. As a consequence the magneto is employed exclusively on vehicles like the motorcycle, while the dynamo is generally used for heavier purposes such as the production of power in factories, lighting plants, etc.

In order to make clear the principle upon which the mechanical generation of electrical energy is based it will be necessary for us to consider the simplest type of mechanical generator that can be conceived. This is shown on page 56. In this simple machine the lines of force are produced between the two poles of a horseshoe magnet and the conductor, which in this case is a loop

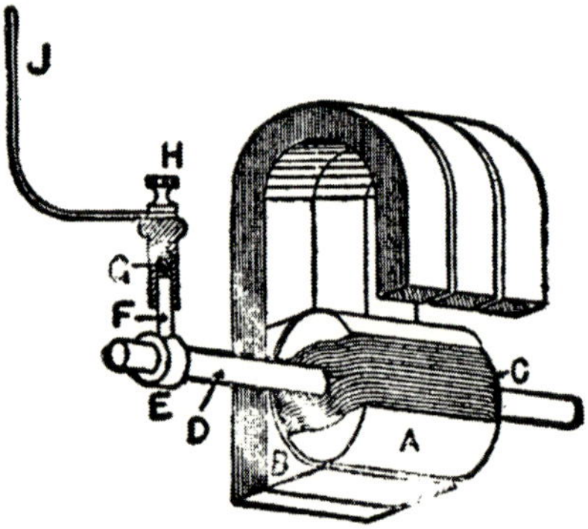

Sectional View of Magneto.

A armature spindle; B field; C wire winding around spindle D; E insulated ring; F carbon brush, pressed down by spring G; H screw terminal; J connecting wire.

of copper wire, is mounted upon a spindle in order that it may be rotated in the field to cut the lines of magnetic influence present between the two pole pieces. Both of the two ends of this loop are connected with the two insulated drums shown upon the shaft. Two metal brushes are bearing upon these drums serving to collect the current and cause it to flow through the external circuit. It can be seen that when the shaft is turned in the direction of the arrow the loop will cut through the lines of magnetic influence and a current will be generated therein.

Make-and-Break Vs. High Tension.

Two forms of magneto ignition are in use, the make-and-break, and the high tension. The former is the older and better understood, but the latter is fast coming into

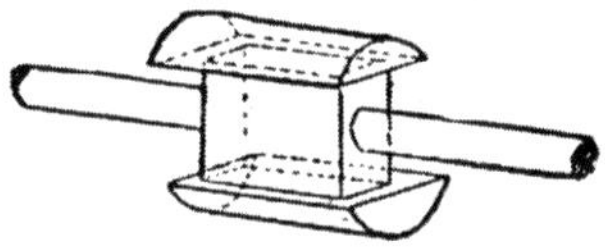

Core of the Armature.

favor on account of having fewer parts, being more simple in operation, and ensuring greater certainty of action. As will be readily understood it is the sole duty of the magneto to produce the current which furnishes the spark by which the explosion is made. This seems to

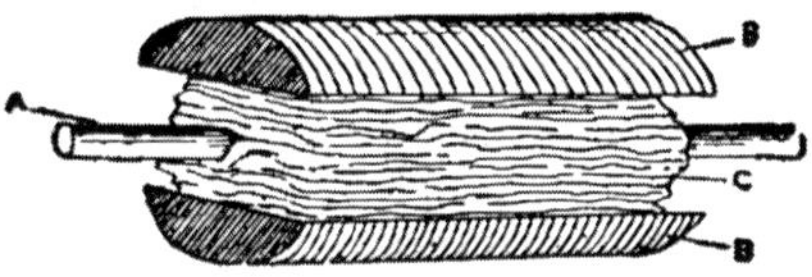

Armature of Magneto.

A armature spindle; B B iron core; C wire coil around spindle.

be provided more surely by the high-tension process than by the make-and-break system. From the magneto a wiring system connects with a spark plug inserted in the head of the cylinder, the contact of the metal at the base of the plug with that of the cylinder furnishing a "ground" which completes the circuit. In connection with the magneto is a contact breaker which interrupts the current at stated intervals, and the result is a spark at the end of the spark plug inside the cylinder, which produces the desired ignition

As it is the purpose of this book to make the reader acquainted with the mechanical operation of the various parts of the motorcycle, we will explain "how" the magneto works, and ignore the theory end of the problem. In the make-and-break system, or as sometimes called "low tension ignition," we have a contact breaker

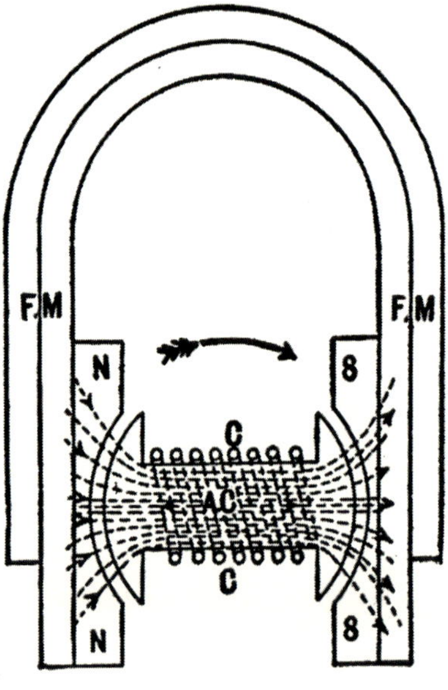

Lines of Magnetic Force Passing Through Armature Core.

The dotted lines show the magnetic field between the poles N and S. A C is the core or armature. F M are the permanent magnets.

inside the cylinder. This is wired directly to the magneto, and by automatic action "breaks" the electric current at stated intervals, causing a spark. This spark is produced by the effort of the current to make a connection between the ends of the wires in the spark plug by jumping across the small gap which separates them. This system works very well in single cylinders, but where there is a multiplicity of cylinders, and necessarily a contact breaker in each, it is evident that all can not be making contact at the same time, and consequently it requires the utmost nicety of adjustment to get them

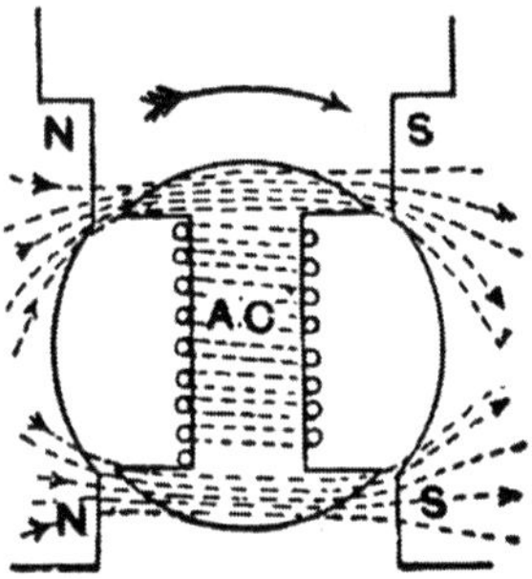

Lines of Magnetic Force Passing From North to South Pole.

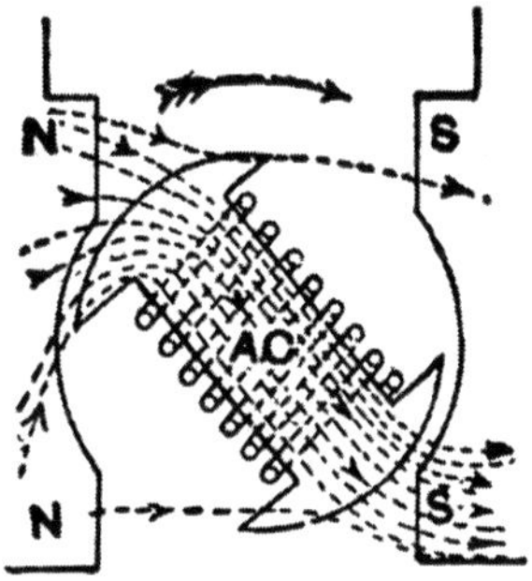

Breaking the Magnetic Field.

in harmonious, effective action as the intervals between the making and breaking of the circuit in each instance are very short, being a mere fraction of a second.

Batteries and Spark Coils.

In the high-tension magneto an induction coil induces a high tension current which of its own volition "jumps" the gap in the sparking plug, thus doing away with the necessity for a contact breaker inside the cylinder. The operation of both systems will be the more readily understood by an inspection of the various illustrations here given, and reference to the explanatory notes in connection with each.

Previous to the introduction of the magneto batteries

were exclusively used to produce ignition, and many motorcycle makers and riders still adhere to them. The "dry" battery is given preference for several reasons. One is its lightness, another its simplicity, and another its ease of manipulation. Every schoolboy who has dipped into the rudiments of electricity understands the construction and operation of the dry battery. Another point cited in its favor is cheapness. Other battery forms may, and are frequently employed, but the dry battery, for the reasons given, is the favorite. This battery is

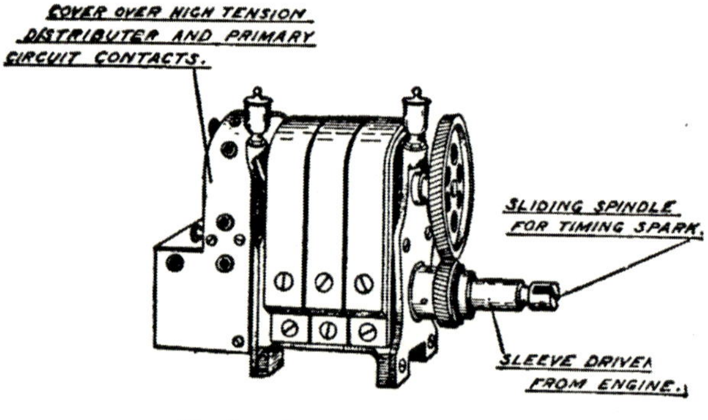

Typical Form of Magneto.

usually utilized in groups of three, this number being sufficient to generate the electric current for any single cylinder machine.

When the battery system is used there must be a spark coil. As it leaves the battery the electric current is of low pressure and large volume. In order to produce a spark, in order to give the current power enough to "jump the gap" at the spark plug and thus ensure ignition, it must be transformed into one of higher pressure and condensed in volume. This is what the spark coil does. As its name implies the coil is a "coil," or rather a series of "coils" of soft iron wires. These are wound in alternate directions and insulated so they do not come into actual contact with one another. The low pressure current from the battery in passing through this coil is transformed, and increased in its energy so as to produce what is called a "fat" spark. It will make the op-

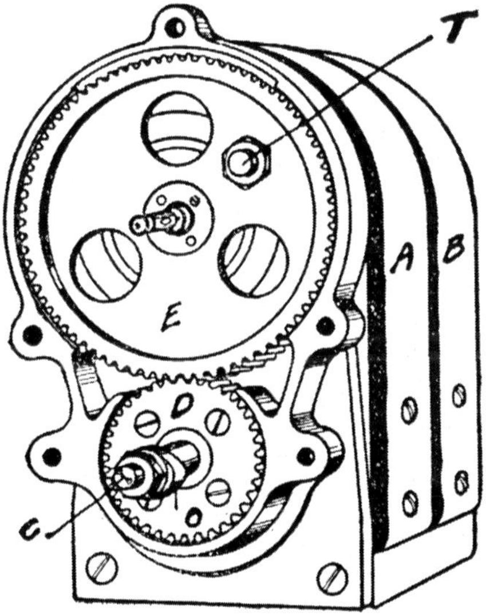

Gear Wheels of High-Tension Magneto.

eration of the coil more readily understood if we take a peep into its interior make-up which is as follows:

The core, or center piece, is a layer of soft iron wire. Around this is wound a couple of layers of thick insulated wire, known as the primary wire, the ends of which are connected within the casing to the terminals. This primary wire is enclosed in an insulator, generally a tube of vulcanite, over which is wound successive layers of a fine high-tension wire, each layer being completely insulated, although the wire itself is continuous. The ends of this high-tension, or secondary wire are also attached to the terminals, within the casing, and the whole liberally coated with some insulating material such as paraffine. As the battery current flows and ceases to flow through the primary wire high-tension currents are induced in the secondary wire. Without this coil arrangement it would be impossible to produce from the battery a spark of sufficient energy to produce ignition.

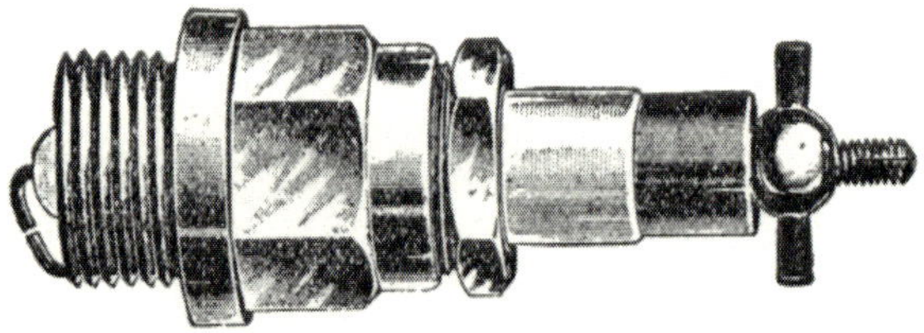

One Form of Spark Plug.

How the Spark Plug Works.

In the top of the cylinder, or at one side near the top, is a small threaded aperture into which the spark plug is screwed. This plug is a comparatively insignificant looking bit of mechanism, but it must operate without hitch or the motorcycle will not work. There are many different makes on the market and some particular advantage or merit is claimed for each, but as in other important parts the basic principle is the same in all. In fact this is one marked characteristic of all gas engine vehicles—they may differ in construction and in methods of applying the motive power, but they are all based on the one principle, and it is this that makes them easy to understand by the man who will take the time to advise himself as to their mechanism. The spark plug is rarely over 2 inches in length, and barely ¾ inch in diameter. The illustration here given is of actual size, and shows the construction. Small as the spark plug is, however, it is fully as important as any other part of the machine. Remove the spark plug, and the connecting link between the fuel and its effect is gone. The motorcycle can not be moved except by laborious pedalling, or shoving.

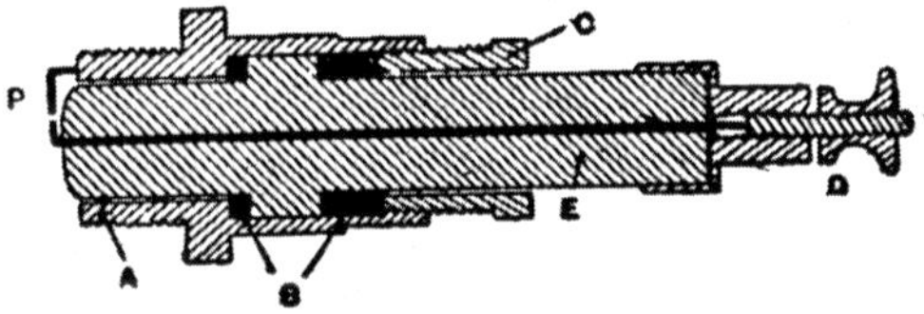

Sectional View of Spark Plug.

A metal body with screw thread to fasten into cylinder hole; B packing; C hollow screw; D terminal for high tension wire from magneto or battery; E insulation; P wire points. The long black line running through the center of the plug is the wire through which the current is conducted; the bent wire, attached to the metal and slightly separated from the one which connects with the magneto or battery wire, leaving a small gap. In attempting to "jump" this gap the current produces a spark. It finally reaches the ground wire, thus completing the circuit. Both wires are platinum pointed, so as to withstand the heat.

While the spark plug is made of metal, usually brass, the interior is of porcelain or mica. This is to secure insulation. From the top of the plug, extending through the interior, and ending in a bent form, is a wire of substantial size. To the end of this wire projecting from the top of the plug the wire from the spark coil or magneto is attached. From the threaded metal part of the plug which screws into the cylinder head, is a duplicate of this wire which almost, but not quite, meets the wire first referred to. There is a gap of about one twenty-fifth of an inch between the ends of these wires. It is by the effort of the electric current to "jump" from the battery or magneto wire to the ground wire that the spark which fires or explodes the gas is produced.

Another Form of Ignition.

One of these days it is probable that another form of ignition will come into use, but it is still in the experi-

Two Forms of Bosch Magnetos.

mental stage. This is known as the hot-air process. Engineers are now working on a plan for compressing a small quantity of air or gaseous mixture in a separate pump, driven from the engine, until it has attained a temperature sufficiently high to secure ignition. In addition to this we have the possibility of what is called the catalytic method. This hinges upon the fact that certain metals, like platinum and palladium when exposed in a spongy state are susceptible to the action of mixed explosive gases. They absorb these gases to a

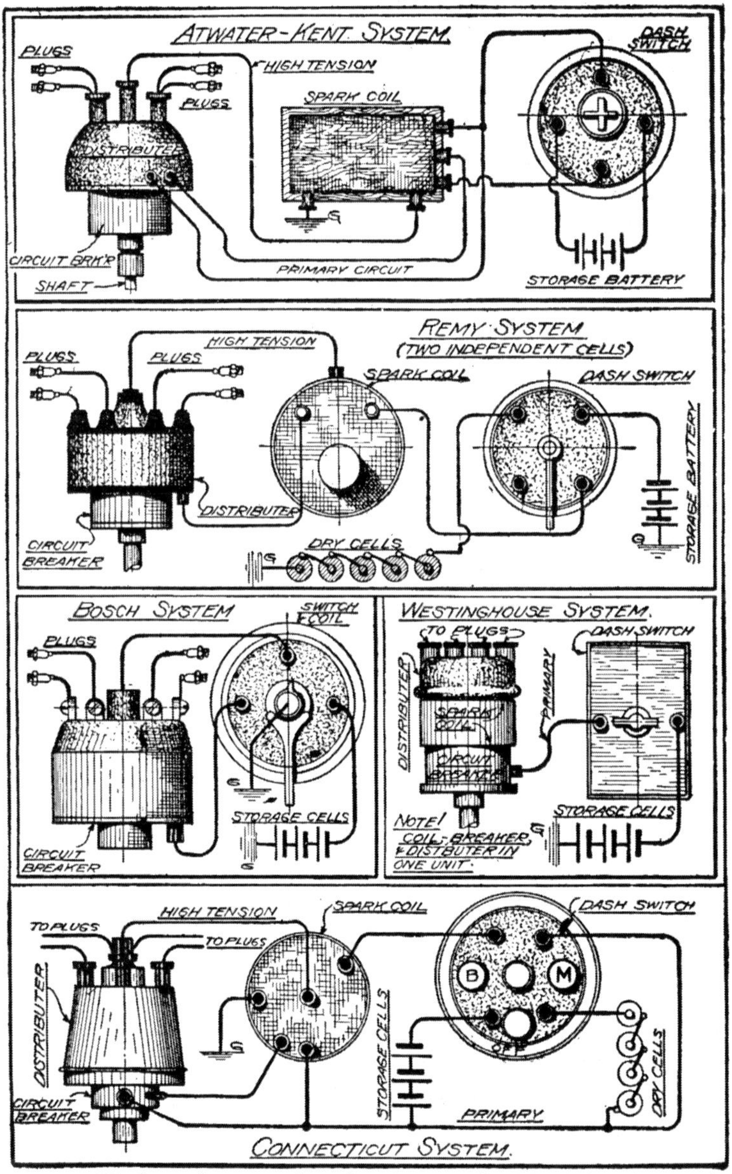

Modern Battery Circuits Commonly Used in Connection With Self-Starting Systems.

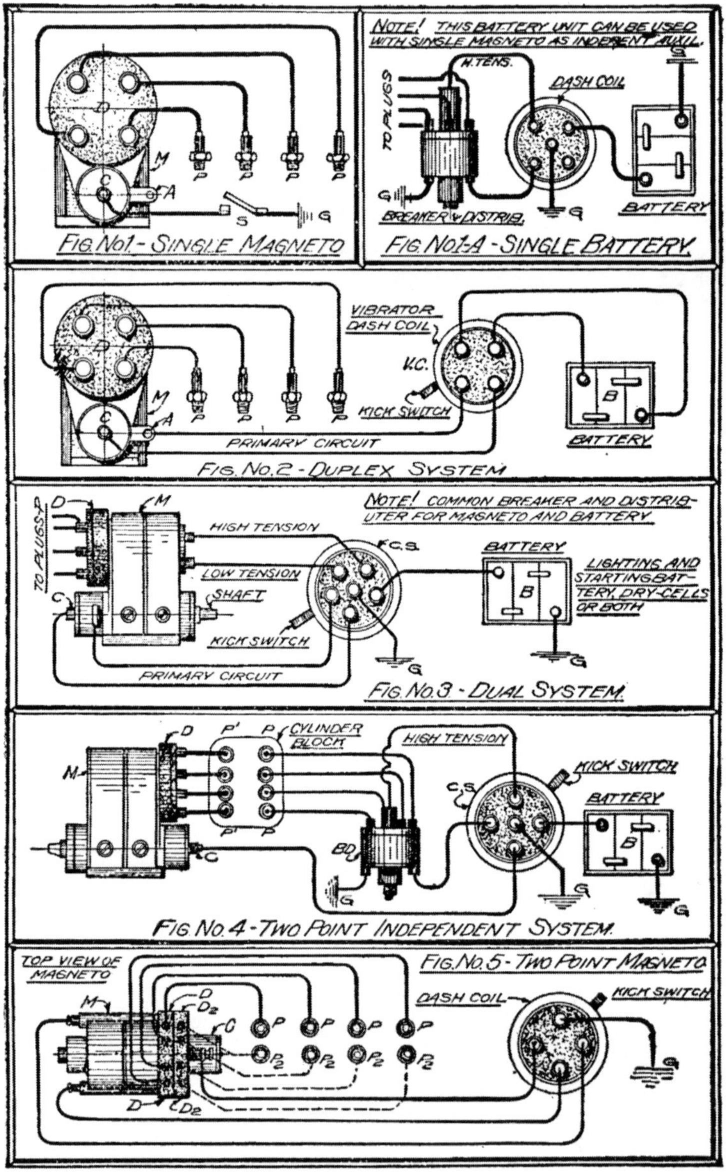

Figs. 1-1A-2-3-4-5.—Showing Different High Tension Magneto Wiring Systems.

certain extent and in doing so become sufficiently hot to explode the gaseous vapor in the cylinder. An illustration of this method is to be had in an ingenious little cigar lighter which is now on the market. This consists of two parallel metal tubes, about two inches in length, brazed together. One contains an asbestos lining, saturated with alcohol. The other is merely a receptacle or holder for the firing point which is a loop of platinum. When this platinum loop is introduced into the alcohol soaked asbestos and withdrawn it is covered with a flame which lasts long enough to ensure a light

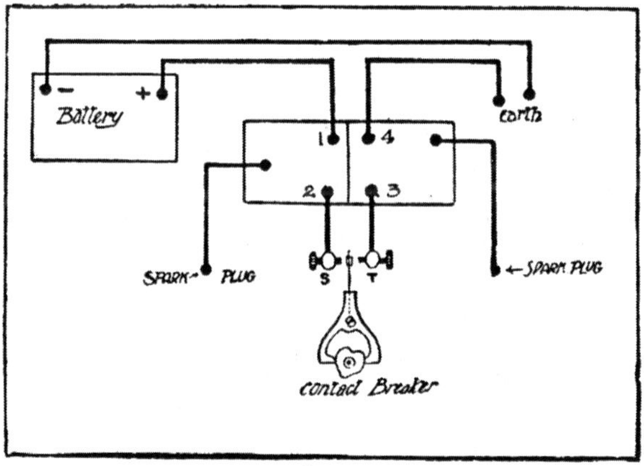

Simple Ignition System for Two Cylinders

for a cigar or lamp. Here we have a possibility for a new method of ignition for motorcycles, but in its present form it is only a possibility.

Methods of Generating Electricity.

There are two methods of generating the electricity required in motorcycle ignition. One is by chemical action, the other by mechanical energy. In the battery system the current is generated by chemicals. This chemical action is obtained in dry batteries by using a solution of salammoniac mixed with carbon dust, sawdust, black oxide of manganese, plaster of paris or similar material. The substance selected is saturated with the "exciting" fluid (salammoniac). Carbon and zinc plates are wired and the exciting mixture, already described, works upon them, the salammoniac dissolving the zinc, forming a free hydrogen gas, which is liberated at the carbon element. This, in brief, is the method by which a current of electricity is generated by chemical action. This dry battery

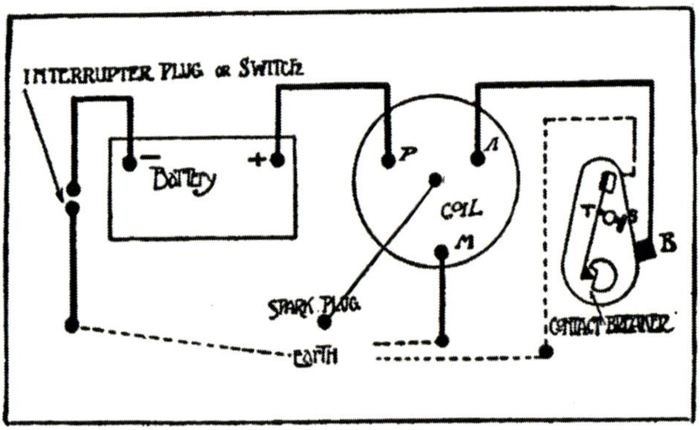

Circuit For One-Cylinder Engine.

system is preferred by motorcyclists to the wet, or galvanic method, as there is no fluid to slop out, no refilling to do, and the outfit is much lighter. When a dry battery cell is exhausted, it is simply a matter of throwing it away and substituting a new one. The price of a dry battery is about 25 cents, making an outlay of 75 cents for the complete three battery equipment required on a motorcycle.

Storage batteries are not used on motorcycles. They are too heavy and cumbersome. As the name indicates, the storage battery is not a generator like the dry battery or the magneto; it does not create electrical energy, merely stores, or houses it; acts as a sort of "go between" between the original source of supply and the application of the current.

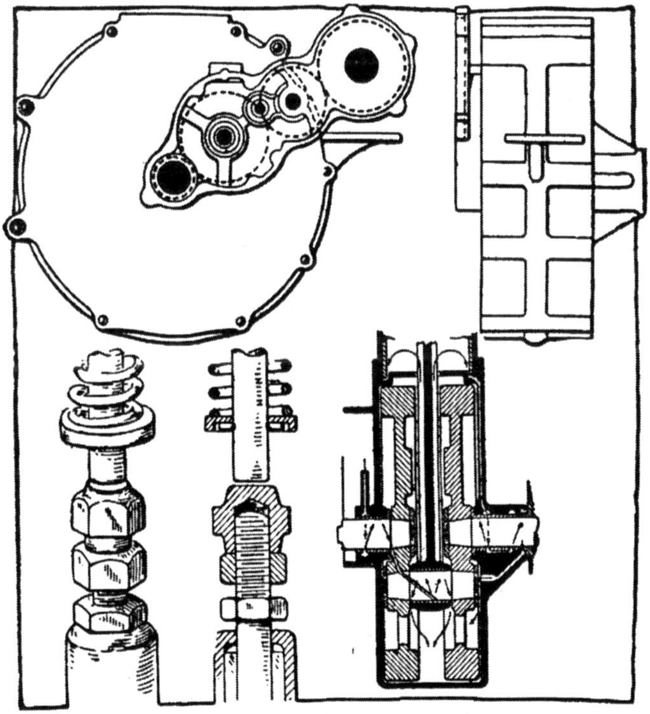

Excelsior Magneto Gearing Adjustable Valve Tappet and Oiling System.

The Magneto Method.

The magneto method of generating an electric current, as previously mentioned, may be described as solely mechanical, no chemicals being employed. The magneto is put in motion and kept in operation by the mechanism of the motorcycle itself. When the engine starts, or the pedals are used, the magneto begins its action. When the machine is not in use the magneto stops automatically. There is another distinct difference between the battery and magneto—the former can deliver only a certain fixed amount of electrical energy, while in the magneto the amount of this energy varies with the speed of the engine. As the speed increases, the current becomes stronger. This is not intended as a plea for giving the magneto the preference, and should not be so

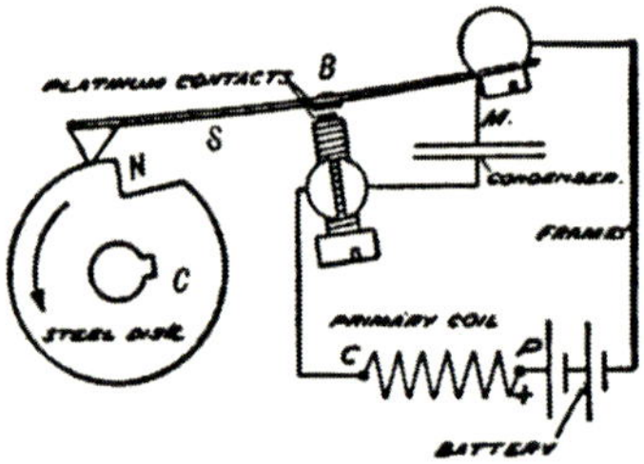

Primary or Battery Circuit.

construed. It is merely a means of illustrating the difference in the working of the two methods. Both, thanks to modern ingenuity, are admirable in action and efficiency, and both have characteristics which appeal to distinct classes of riders, according as tastes and ideas may vary.

Magnetos and Dynamos.

In the magneto used on the motorcycle, the magnet is always a permanent one, as this form is the lightest and most compact. Between the poles, the extreme lower ends of this magnet, what is known as a magnetic zone or field, exists. If this "field" is cut up, or interrupted often enough, an electric current is produced, its intensity increasing proportionately with the number of

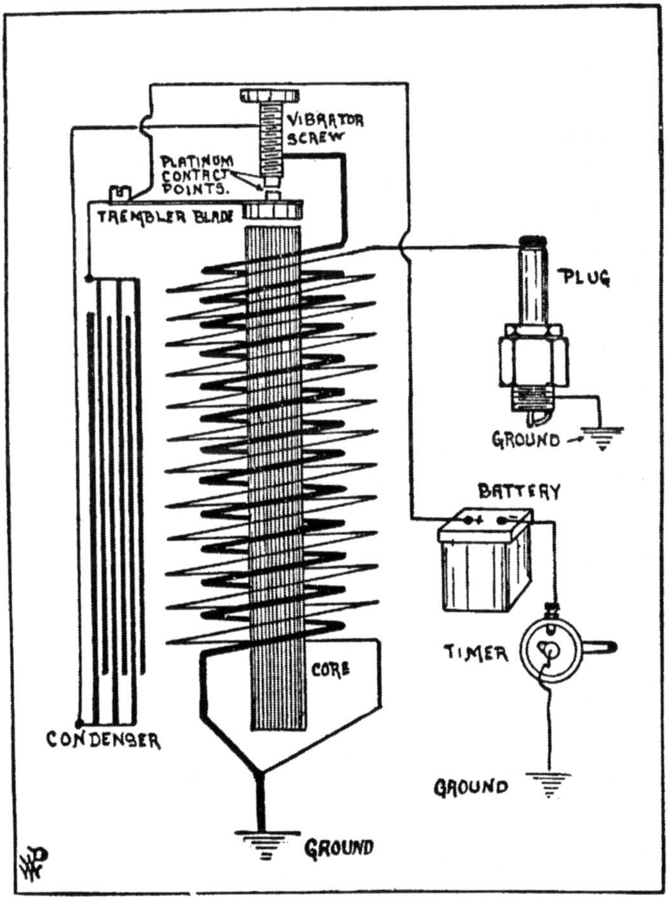

Working of the Induction Coil.

"interruptions." When the motorcycle is set in motion, this interruption of the magnetic field, caused by a spe-specially constructed part of the magneto, begins increasing as the speed of the motor is heightened. This is mechanical electricity.

On larger vehicles, such as automobiles, for instance, dynamos are used. These perform the same service as magnetos, but produce the electricity in a different manner. In the dynamo, the permanent magnet is discarded and an electro-magneto substituted. In this latter the magnetic field is induced as wanted by the intervention of a special winding of wire. If an expert should be asked to explain the difference he would say: "In a

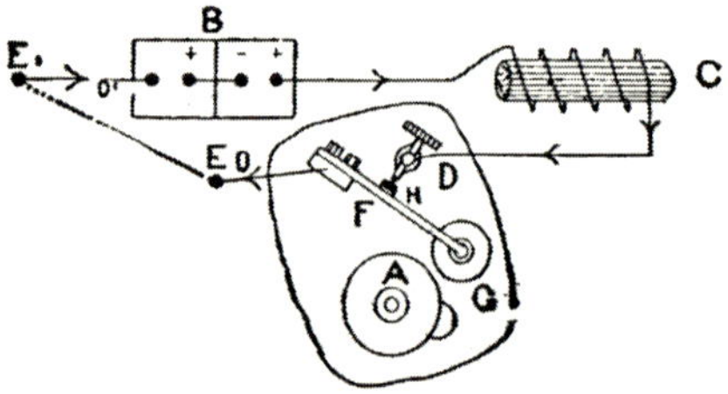

Make-and-Break Wiring.

B battery or accumulator; C coil; E connecting wires; A D F G parts of contact maker and breaker already explained.

dynamo the electric current is created by electro-magnetic production, while in a magneto it is produced by magneto-electric induction." Clear, isn't it? And yet it may be understood when we stop to remember that the dynamo actually creates a magnetic "field," while the magneto merely utilizes one that is already in existence. Dynamos are never used in motorcycles; like the storage batteries they are too heavy and cumbersome.

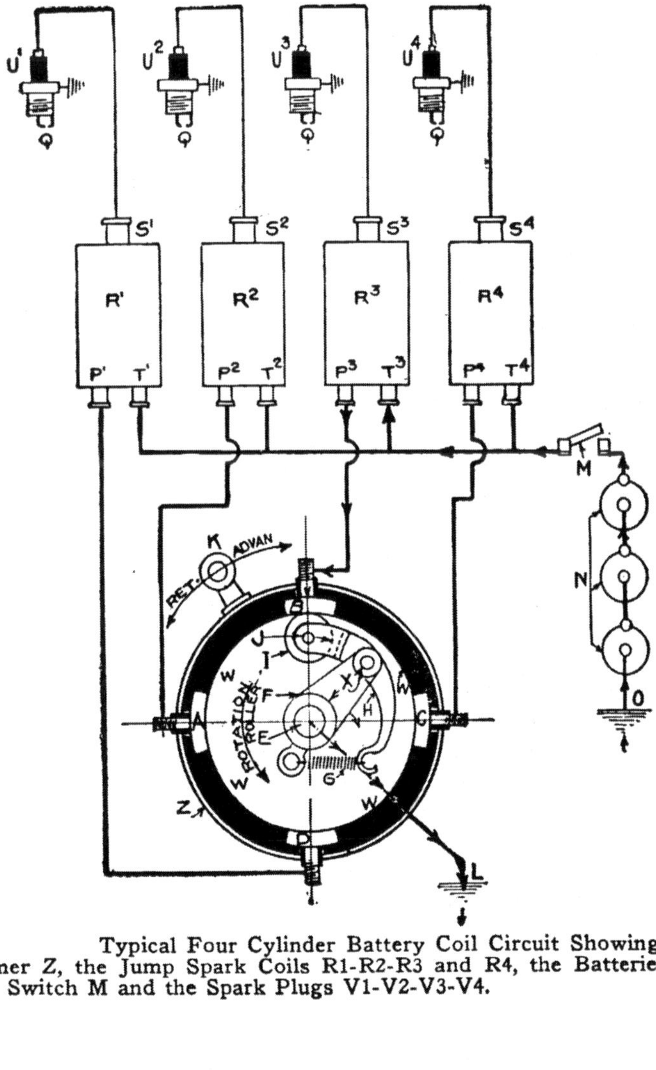

Typical Four Cylinder Battery Coil Circuit Showing the Timer Z, the Jump Spark Coils R1-R2-R3 and R4, the Batteries N, the Switch M and the Spark Plugs V1-V2-V3-V4.

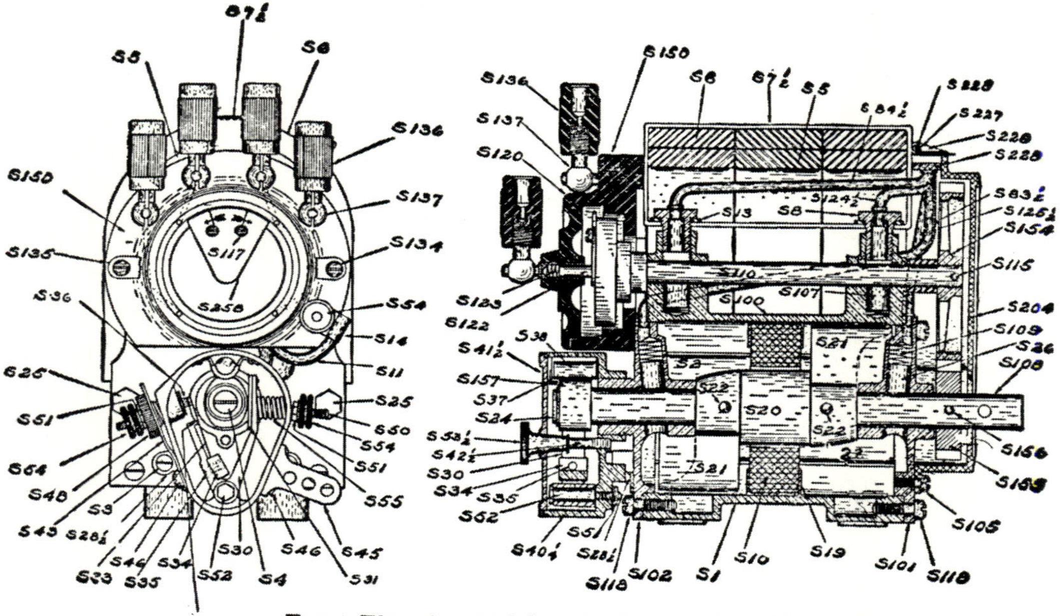

Front Elevation and Longitudinal Section Through Remy Magneto.

CHAPTER VI.

THE MOTORCYCLE ENGINE.

With the explosion of the gas in the cylinder we get the inception of the motive power, the turning of which into actual propulsion depends upon the action of the cylinder mechanism. Most motorcycles are of the one cylinder type, but there is a tendency to multiply the number of cylinders with the purpose of gaining more power and speed. This, however, does not affect an illustration of cylinder action, as they all work on the same plan. The engine in general use today is of what is called the four-cycle type. This means that the piston stroke, from start to finish, performs four different and distinct functions. These are fully explained in the accompanying illustrations.

The first duty of the engine on its initial downward stroke is to open the inlet valve and "suck" in a charge of gas. This is called the suction stroke, and is cycle No. 1.

It next compresses this gas by an upward movement, so as to secure the greatest possible force or energy when the explosion takes place. This is the compression stroke, or cycle No. 2.

Then comes the explosion, or power stroke by which motion is communicated to the crank shaft. This is cycle No. 3.

Having exploded the gas when the piston head is close to the top of the cylinder the piston naturally is driven downward again. This leaves the space above the pis-

Stroke No. 1.

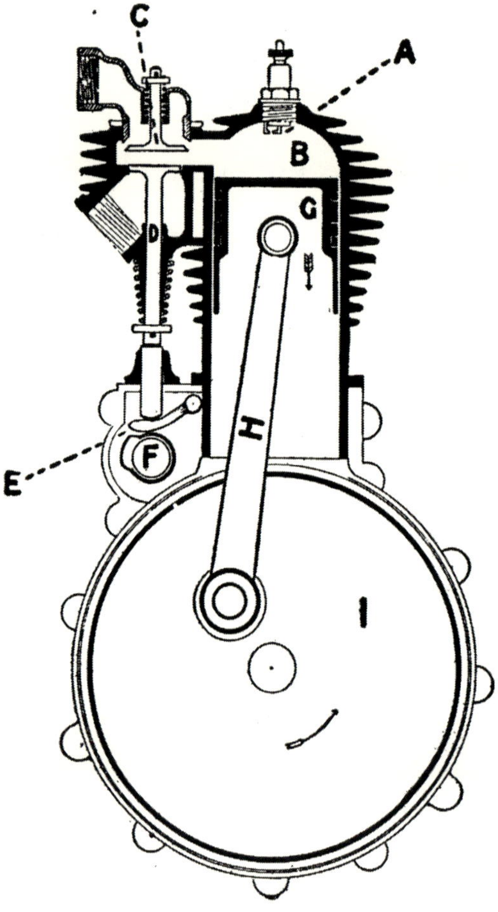

Action of Armac Four-Cycle Motor.

A is the spark plug; B the combustion chamber in the cylinder; C the intake valve through which the gas is admitted; D the exhaust valve; E the cam which opens and closes valve D mechanically; F head of shaft which operates the cam E; G piston; H connecting rod; I crank case.

Here we have the engine starting on its initial or suction stroke. Note that the suction caused by the downward movement of the piston has drawn open the intake valve C, and through this the gasolene vapor enters the combustion chamber B.

Stroke No. 2.

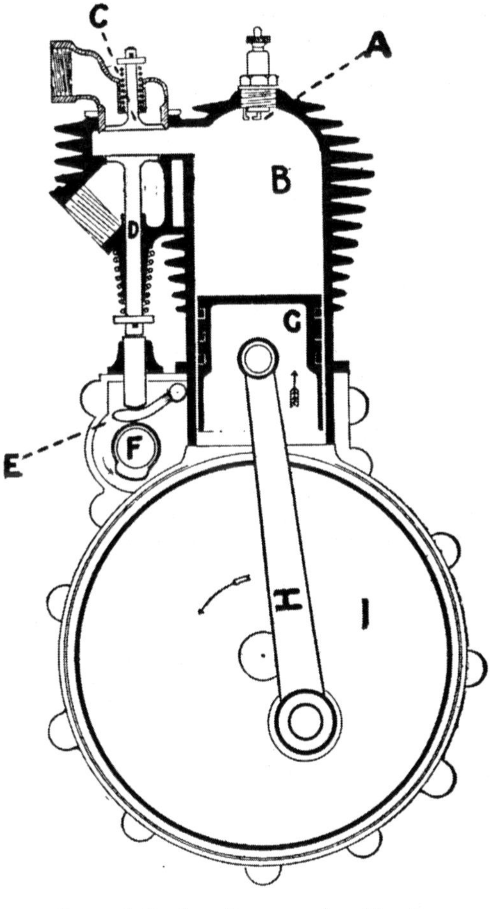

Second Cycle—Compression Stroke.

This is the beginning of what is known as the compression stroke. Both intake valve C, and exhaust valve D are closed, and as the piston moves upward it compresses the charge of gasolene vapor in the chamber B.

ton head filled with the exhaust gases which must be expelled before a new charge of explosive can be taken in. This is accomplished by the exhaust stroke which opens the exhaust valves and allows the burnt gases to escape.

Here again a study of the illustrations is essential to a thorough understanding of the operation of the engine movement. As will be seen the cylinder contains a piston head which fits snugly so as to be virtually air tight. To aid in this both the interior of the cylinder, and the edges of the piston which come into contact with it are ground to a velvet-like smoothness. This not only produces an air-tight chamber in the cylinder, but reduces the friction to the minimum so that the greatest possible amount of propulsive energy may be created with the smallest possible expenditure of power. The smoothness with which some of these pistons work is little short of miraculous. The writer has, for instance, seen a piston set in operation by a gentle hand movement, without the use of explosive of any kind, run for thirty minutes without stopping. This illustrates the exactness of detail in mechanical construction which prevails in the modern motorcycle as the case cited is one in which an ordinary machine, taken out of stock, was used. It was, of course, suspended, so that the wheels did not come in contact with the ground, and by this the friction was greatly reduced. Still, it was a wonderful performance.

Despite the smoothness to which the cylinder interior and the piston head are ground it would be impossible to secure the necessary "snugness of fit" were it not for the piston head rings. These are three circlets of fine grey cast iron, with a slot cut in each as shown in the illustration. This admits of the rings compressing or expanding as occasion demands. The face of the piston

Stroke No. 3.

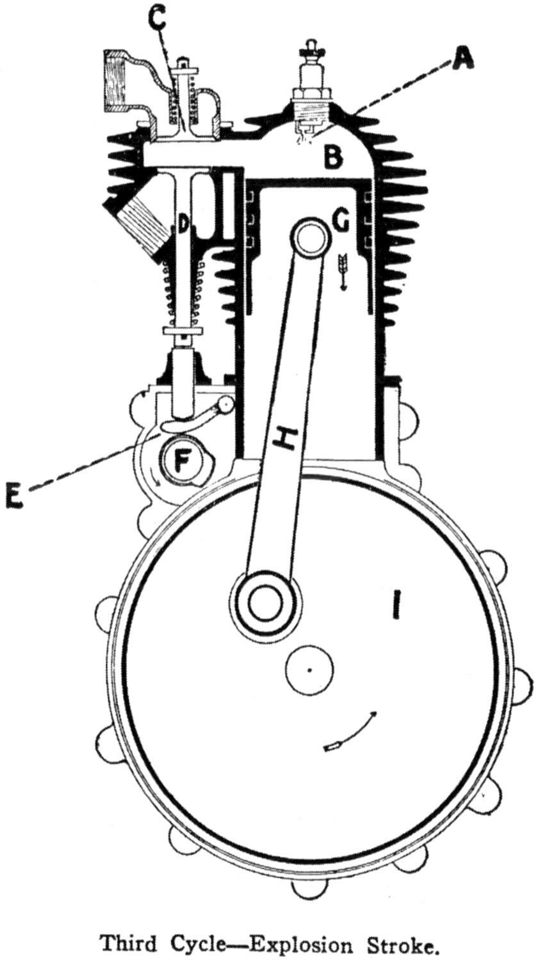

Third Cycle—Explosion Stroke.

Here we have cycle No. 3, or the explosion stroke. Both valves C and D are still closed, and the gasolene vapor has been compressed into the small space shown in B. At this junction a spark is emitted from A. This explodes the gasolene vapor and drives the piston downward again.

is grooved to admit of these rings being "sprung on" at a short distance apart. In their original form they are slightly larger than the bore of the cylinder but the slot which has been cut in them allows of their being compressed so as to fit. This fit is necessarily a tight one, but the smoothness of the surface, the thorough lubrication they receive, and the nature of the material of which they are made, reduces friction to the lowest possible point. In placing the rings on the piston care must be taken that the slots are not in line—that they do not come directly one above another. If placed in this way an opening is left through which gas will escape, causing a decided loss in explosive power. In placing the rings the slots—let us call them openings in this one instance for sake of clearness—should come so that the solid part of one ring will be directly over the opening in the one below it, and so on.

Piston Rod Connections.

The connecting rod, running from the piston head to the crank shaft, connects with a system of gears which operate various valves mechanically, thus ensuring perfect harmony and regularity in action. That this latter is an all-important feature may be understood when we know that there are from 300 to 500 explosions in the cylinder chamber every minute, and that consequently there must be exactness to the smallest fraction of a second, or serious trouble will ensue. This gear action, for instance, controls the exhaust valve. (See illustrations). Suppose that by some mishap an explosion should occur when the exhaust valve was open. Its force would be lost. On the other hand if the explosion should occur when the intake valve was open the merest tyro knows that a backfire which might possibly explode the gasolene tank would ensue. But such in-

Stroke No. 4.

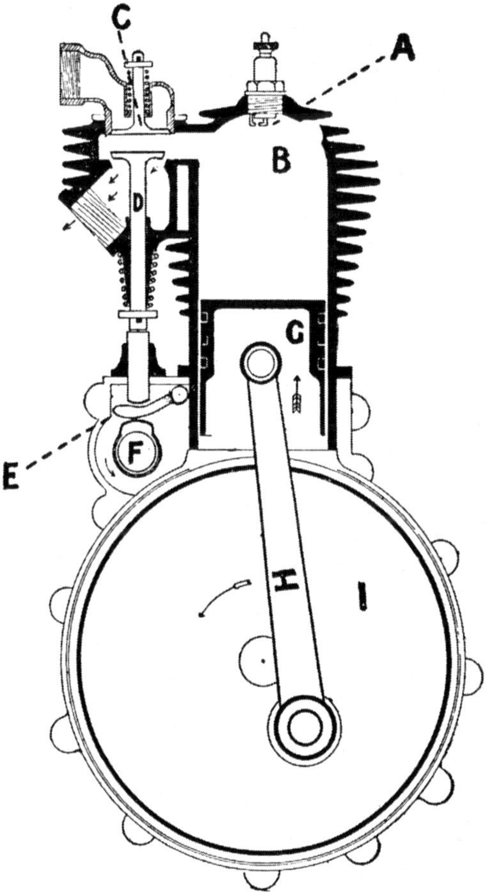

Fourth Cycle—Exhaust Stroke.

In this cut the last cycle, or stroke of the piston is shown. It is called the exhaust stroke. By the action of the gears inside the crank case the shaft F has been moved until the cam or eccentric which is shown at the left of the shaft in figure 3, is brought directly under the lever E. This forces E upward raising the exhaust valve D so that the burnt gas may escape, and chamber B cleared for a new charge. The piston in the meantime, aided by the revolutions of the fly-wheel has begun to work upward and this tends to drive off the burnt gas through the only opening by which it can escape, that made by the raising of the valve D.

stances are extremely rare because the workings of the motorcycle, once properly regulated, are mathematically correct.

The mere force of the explosion in itself would not be sufficient to ensure continuous propulsion especially in rough going. Hence, to ensure this continuous ac-

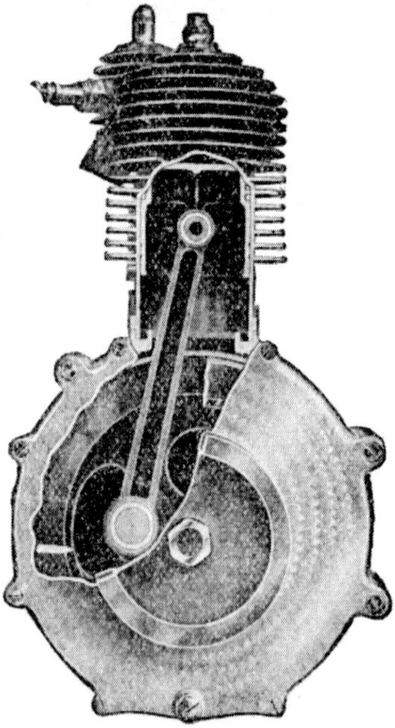

Pulley Side of Motor.

tion, the connecting rod is attached to a fly wheel of sufficient weight, when once set in motion, to overcome the dead center proclivity and assist the piston in returning to the position where it may receive the impetus of another explosion.

Starting the Engine.

Now let us return to the operation of the engine. You wish to start your motorcycle. This can not be done unless there is compression in the cylinder. Ordinarily this is obtained by pedalling, but in at least one modern machine the pedals are left off, and a cranking device,

Connecting Rod Attached to Fly Wheel.

similar to that used on the automobile, substituted. In either case the purpose is to set the machinery in motion so that the intake valve will open and let in the gas. Next compression will ensue with following spark and explosion and finally continuous action. Note in stroke 1 the position of the intake valve C. When the piston descends the seat of this valve is drawn downwards by suction, and a charge of gas is sucked in. Now in the

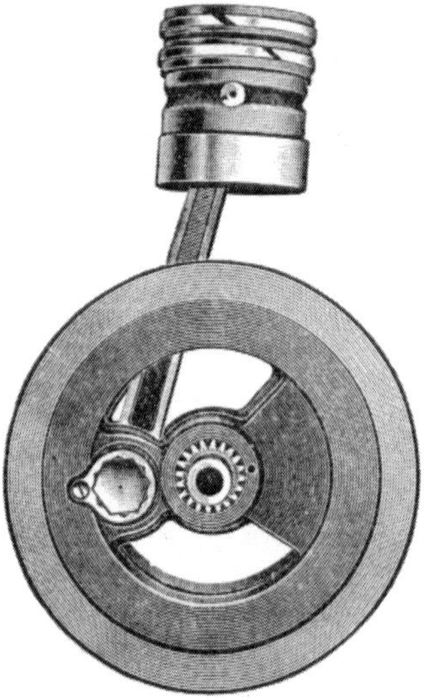

Piston and Fly-wheel Connection.

compression stroke, as shown in cut No. 2 we see that the intake, as well as the exhaust valves are closed, the seats of both being snugly fitted against their bearings. The piston works back in an upward movement until the gas is properly compressed. See cut No. 3. At this juncture the spark comes and the explosion occurs. Back goes the piston to the bottom of the cylinder, as shown in cut No. 4, the gear action at the same moment raising the exhaust valve so the burnt gas may escape. In some motors, especially multi-cylinders, both intake and exhaust valves are operated mechanically, but as a rule

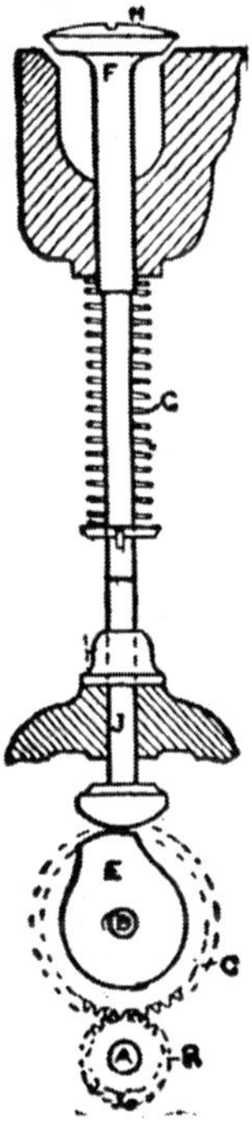

Exhaust Valve.

It is shown enlarged, and opening, the valve H being pushed up from the seating to allow the burnt gas to exhaust or escape. A is the crank shaft. B is the pinion on this shaft. C is another pinion attached to the cam E which in turn is revolved by the shaft D. In its revolutions the cam E raises the plunger J which in turn acts on the valve stem F lifting the valve away from its seating. G is the spring which draws the valve back (closes it) when the pressure of the cam is taken off.

the intake valve is worked automatically, while the exhaust is worked by the gear action as shown in the illustrations. Once the cylinder mechanism is in operation the rider may stop his pedalling or cranking, and let the engine continue the work.

The Exhaust System.

An important feature directly connected with the engine is the exhaust. It is through this that the burnt gas escapes when released by the exhaust valve. From the valve it passes through a pipe or tube to an opening from which it escapes into the open air. If this gas is allowed to escape without opportunity for expansion, a lot of noise, echoes of the explosions, is heard. To smother this noise a "muffler" is used. This is merely an enlarged chamber or receptacle placed at the exhaust opening. It has a number of passages in winding through which the burnt gas gets an opportunity to expand before passing into the atmosphere. In doing this the noise is lost, and all there is left is a mild, almost unnoticeable

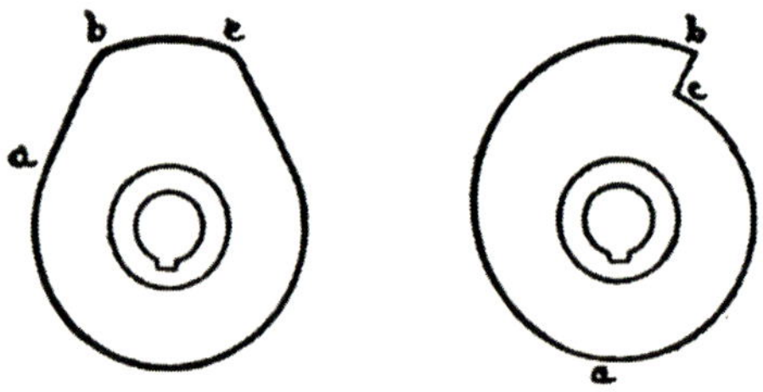

Two Forms of Lifting Cams.

At the left is the ordinary form of cam used in lifting valves. The contour from A to B constitutes the lifting stroke; from B and C the valve is held open, this being the highest part of the eccentric; from C on the valve closes.

smell. There is nothing mechanical about the operation of the exhaust, aside from the valve itself. It is merely a matter of the burnt gas seeking an outlet with the least possible resistance.

If the muffler could be kept closed all the time the noise to which many people strongly object would be eliminated. This, however, is impracticable, as the retention of the burnt gas in the muffler tends to create a back pressure which counteracts the energy of the engine and puts a severe tax upon its force. When the muffler is open the used gas escapes freely as fast as released from the explosion chamber, but when the muffler is closed its escape is greatly retarded. Consequently it accumulates, not only in the muffler itself but in the tube leading to it until at last the back pressure, as it is called, reaches the cylinder. In order to overcome this it is necessary for the rider to keep the muffler open

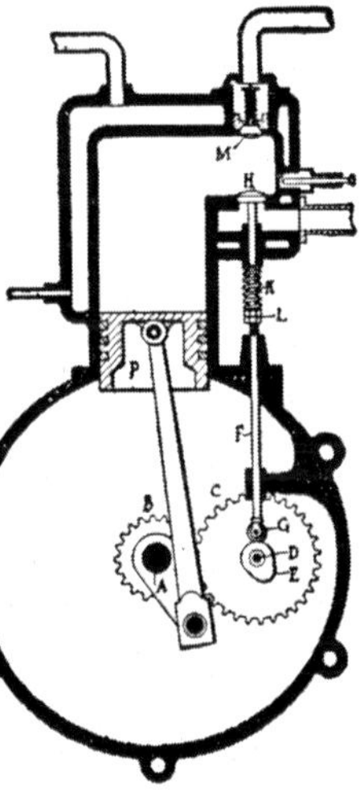

Location of Cam Shaft.

A crank shaft; B C gears; C is double the size of B and hence revolves at half the speed of B; the valve to be opened is H; this is done when the cam E revolves and rises against G on the bottom of the rod F, which in turn lifts the valve stem; M is the intake valve.

for a part of the time. The problem of devising an exhaust system that will be practically noiseless and at the same time do away with the back pressure is receiving the attention of brainy men and it is probable that a satisfactory solution will be reached in the near future. In the meantime we must accept the situation as it is and tolerate the present exhaust system as the one great drawback to the pleasures of motorcycling.

There is one thing the rider should keep continually in mind—When his machine is in operation the muffler and exhaust pipe are bound to be intensely hot. If he would avoid severe burns he must not touch either with his bare fingers.

Cooling of the Cylinders.

As a matter of necessity all motorcycle engine cylinders are of the air-cooled type. Water-cooled systems such as are employed on automobiles are out of the question owing to their bulk and weight. Rapid explo-

Two Forms of Air-cooled Cylinders.

sions of highly compressed gas inside the cylinder naturally generate an intense heat and to keep this heat reduced as much as possible a surface much larger than the mere exterior area of the cylinder itself is exposed to the air. This increase of surface exposure is obtained by casting a series of flanges on the outside of the cylinder body as shown in the illustration. When the machine is moving at a brisk pace the air strikes in between these flanges or projections and, as the surface thus presented is larger than that of the actual cylinder itself the cooling effect of the air is naturally increased. Some idea of what this cooling effect is, even on a hot day, may be had from a simple experiment. The next time you are riding when the temperature is well up toward the 100° mark, let one hand drop for a few seconds by the side

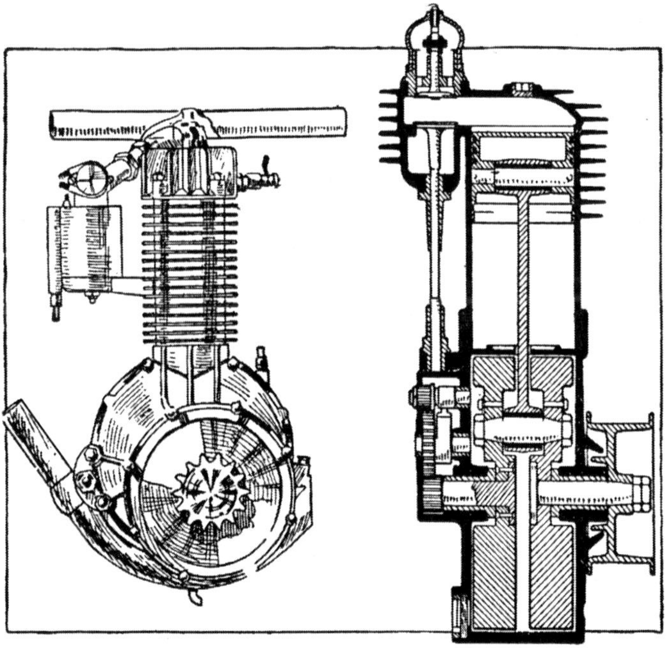

Thor Motor at Left, Yale at Right.

of your machine, with either palm or back part extended forward in the direction you are moving. There will be a very perceptible lowering of the temperature on the part against which the air strikes, sometimes so pronounced as to occasion a slight chill. This is caused by the rapid movement of the machine against the wind, and not from any actual lowering of the temperature itself. It is on this well-understood principle of physics that the air-cooled cylinder is introduced on motorcycles. The same effect noticeable on your hand is produced on the cylinder but in a much greater degree owing to the increased surface obtained by the use of the

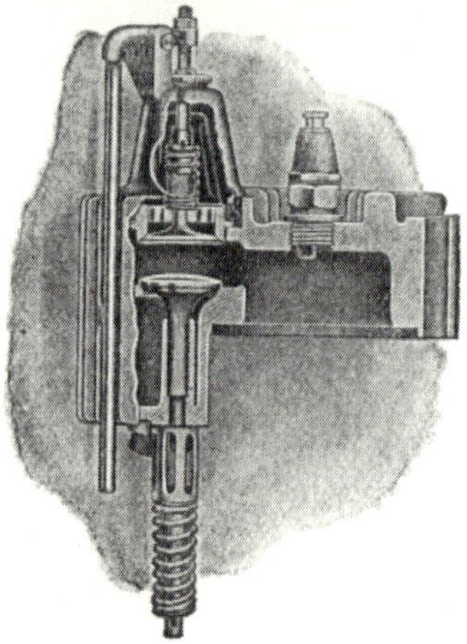

Intake and Exhaust Valves.

exterior flanges. If your machine is standing still no air effect is noticeable. Hence it is a good thing to shut off the engine action whenever you come to a stop as this tends to keep the heat down at a time when there is no excuse for generating it.

Multiplicity of Cylinders.

Motorcycles are now made with one, two and four cylinder motors. That there is an advantage in the multiple cylinder so far as securing additional power is concerned is unquestionable, as it is a plain mathematical proposition that two cylinders of 3½ horse power each will generate just double the power of one, and the ratio increases as additional cylinders are added. Besides this there is more steadiness of what is known as "impact," and therefore less vibration which in the past has been an objection in the one cylinder machine. This, however, has been overcome by modern ingenuity in the hanging

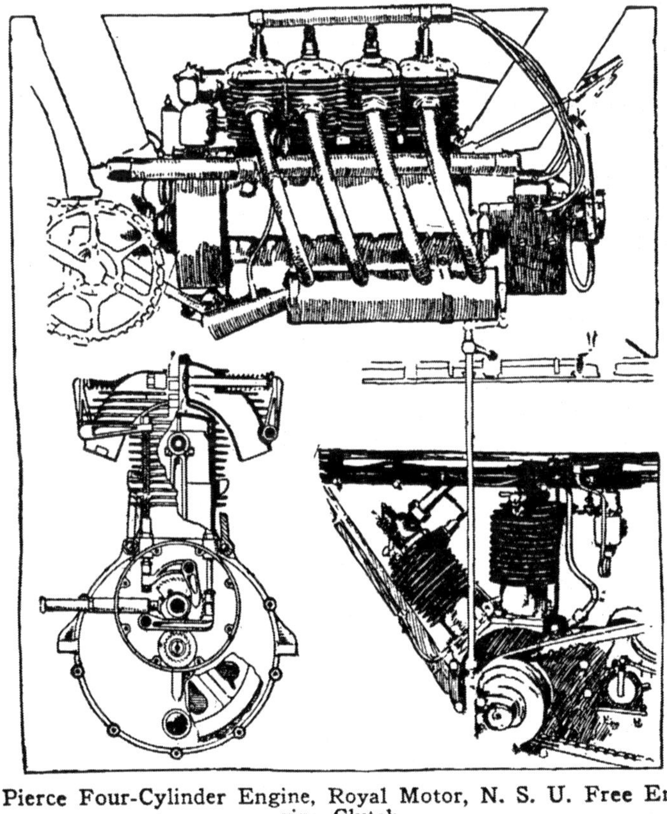

Pierce Four-Cylinder Engine, Royal Motor, N. S. U. Free Engine Clutch.

of the engine and the one cylinder type now rides about as smooth as the two or four cylinder.

When the twin, or two-cylinder, form is employed the cylinders, as shown in the accompanying illustration, are set at an angle of from 45° to 60°, and even 90°, with the connecting rods both coupled to the one crank shaft. The interior operation in each is exactly the same as in the one cylinder outfit, but the timing of the explosions is so arranged that when one piston is making the suction stroke and taking in gas, the other is completing the compression stroke. By this arrangement, and the "quartering" connections with the crank shaft two separate and distinct applications of power are secured at the point (the crank shaft) from which it is transmitted to the rear, or propelling wheel. This form is usually known as the V, from the fact that the placing of the cylinders suggests the letter V.

As a general proposition the cylinders are placed at an angle of about 45°, and some objection is made to this on the ground that, as both are air cooled, the rear cylinder naturally does not receive the same amount of cooling current as the front one, this latter cutting off and absorbing most of the wind.

Placing of Four Cylinders.

In the four-cylinder type the cylinders are placed close together in a straight row (see illustration) and for this reason, if there is anything in the objection of lack of opportunity for air cooling as made in the case of the two-cylinder machine it is intensified in the four-cylinder. As in the case of the two-cylinder motor the connecting rods of the four cylinders are attached to the crank shaft at varying angles so as to secure continuity of action in accordance with the succession of explosions in the various cylinders. Each cylinder acts independently of its mates, there being a minute fraction of a second difference in the time of explosion in each, and in this way the various pistons are at the same instant each performing a slightly different function. While 1 and 4 for instance, are all the way down, 2 and 3 are all the way up. This is a homely explanation, but reference to the cut of the crank shaft will make the meaning plainer than if more mechanical terms were used.

It is frequently said that a novice should not take up a multi-cylinder machine as it is too complicated. There is something in this, and as a rule the one cylinder is best for the beginner as it is easier to operate and understand. At the same time it must be admitted that the multi-cylinder machine has one distinct advantage. The cylinders work independently, each performing all

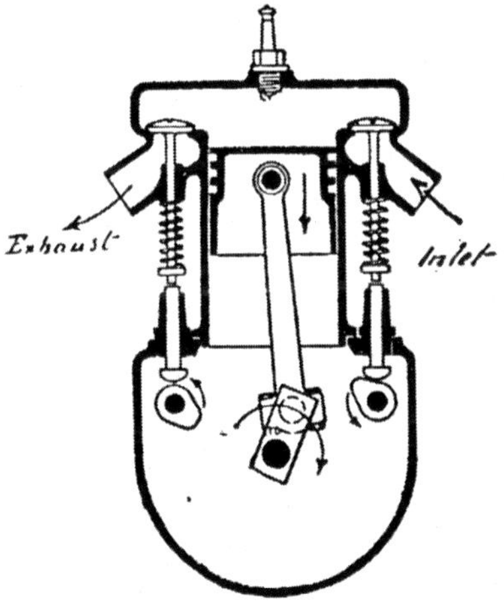

Position of Valves, Valve Cams, Piston and Crank When Spark Has Just Occurred.

the functions required for propulsion. If a man has a two-cylinder machine and one cylinder breaks down, he can operate with the second cylinder, and get home in comfort. The same holds good as to the four-cylinder rig, but in even greater proportion, as in this latter type accidents might happen to three of the cylinders at the same time (something rarely heard of), and yet the rider would have left what, under the circumstances, would be equivalent to a one-cylinder outfit.

A NOVELTY IN MOTORCYCLES.

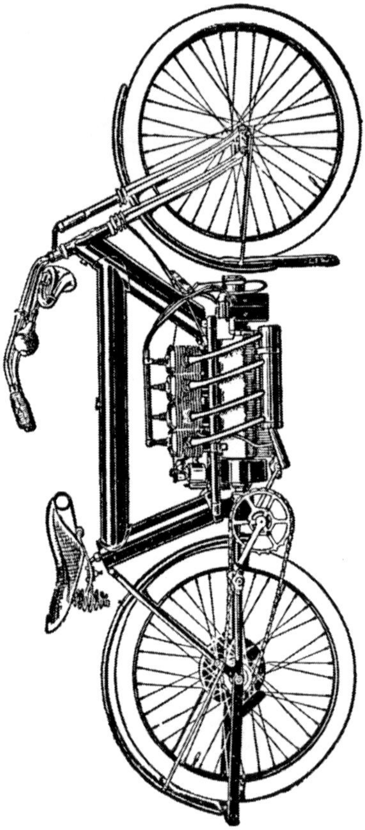

Pierce Four-Cylinder Machine.

	Cylinder No. 1	Cylinder No. 2	Cylinder No. 3	Cylinder No. 4
Firing Order: 1-2-4-3	Firing.	Compression.	Exhausting.	Intaking.
Firing Order: 1-3-4-2	Firing.	Exhausting.	Compression.	Intaking.

Crank Arrangement for 4 Cylinders.

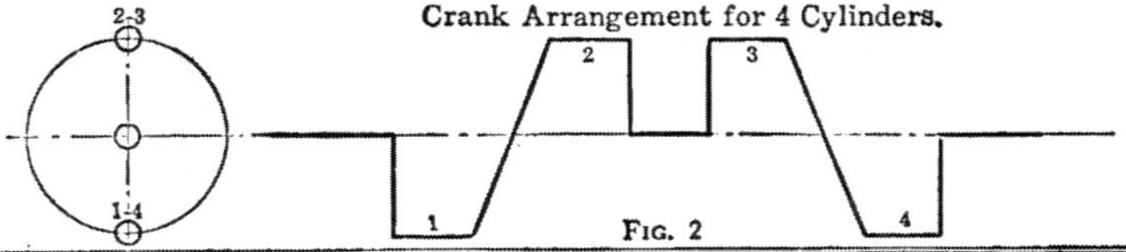

FIG. 2

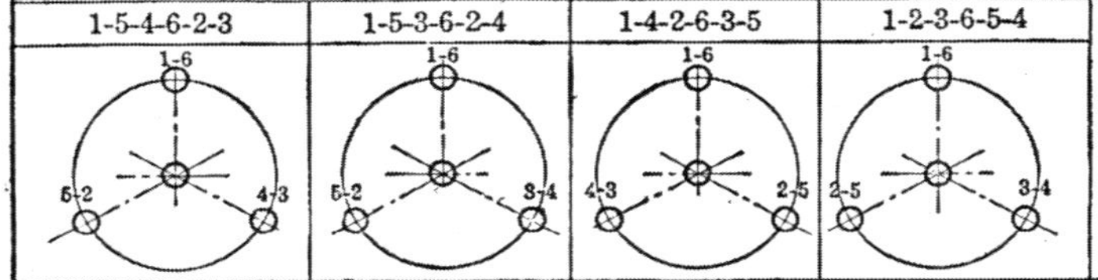

Crank Arrangement

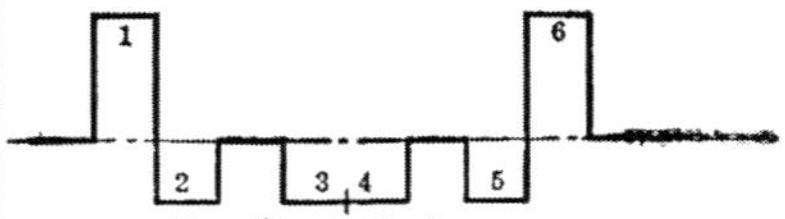

For Six Cylinders.

Firing Order.	Positions of Pistons in Six Cylinders.					
	Cylinder No. 1	Cylinder No. 2	Cylinder No. 3	Cylinder No. 4	Cylinder No. 5	Cylinder No. 6
1-5-4-6-2-3	Top C.L. Begining to Fire	⅓ from Top C.L. on Compression	⅓ down from Top C.L. on Intake	⅓ down from Top C.L. on Intake	⅓ from Top C.L. on Compression	Top C.L. Ex. Closing In. Opening
1-5-3-6-2-4	"	"	"	"	"	"
1-4-2-6-3-5	"	⅓ down from Top C.L. on Intake	⅓ from Top C.L. on Compression	⅓ from Top C.L. on Compression	⅓ down from C.L. on Intake	"
1-2-3-6-5-4	"	⅓ from Top C.L. on Compression	⅓ down from Top C.L. on Intake	⅓ down from Top C.L. on Intake	⅓ from Top C.L. on Compression	"

Firing Order Chart for Four and Six Cylinder Motors.

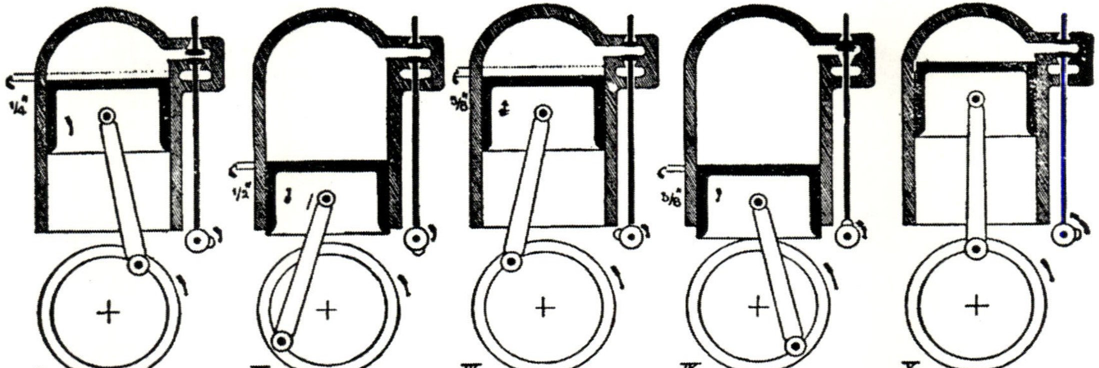

I
SUCTION STROKE.
Inlet valve *opens* ¼ in. *after* piston has started to descend.

II
COMPRESSION STROKE.
Inlet valve *closes* ½ in. *after* piston has commenced to ascend.

III
COMPRESSION STROKE
Spark contact is made ⅜ in. *before* piston has reached dead center, with ignition fully advanced, after which downward power stroke takes place.

IV
EXHAUST STROKE.
Exhaust valve *opens* ⅜ in. *before* piston commences to ascend.

V
EXHAUST STROKE.
Exhaust valve *closes* when piston has reached the dead center.

The dotted lines in all cases denote the dead center at each end of the stroke.

The times of opening and closing the inlet valves only refer to those valves which are mechanically operated.

In this diagram the positions of the piston are shown at the time when the inlet valve opens, when it closes, when the sparking should take place, the exhaust valve opens, and when it closes. Different makers adopt different timings, but above is about the average of 14 well-known makers.

The timing of the spark contact refers only to high tension ignition. When the contact is broken inside the cylinder, as in the low tension magneto system, it should be broken almost immediately on the dead center.

N.B.—This diagram shows automatically operated inlet valves. The timing will be the same with mechanically operated inlet valves.

The Four Stroke Cycle.

CHAPTER VII.

TIMING—WHAT IT IS—HOW DONE.

Unless the explosions of gas in the cylinder chamber occur at exact, regular intervals, the movements of the motor will be uncertain and unsatisfactory. To insure this vitally important regularity in the recurrence of the explosions, a mechanical process known as "timing" is resorted to. This process, applied in a different way,

Dow Magneto With Cap of Timer Removed.

also brings into harmonious action with the explosions the movements of the various valves, so that the functions of each are performed at the proper instant. It is by this means only that the desired results are to be obtained with a synchronical certainty. There is no part of the motor vehicle, whether cycle or automobile, that has caused more brain-fag to inventors than the development of a satisfactory timing system, and that it is in practical operation is a high compliment to the ingenuity of the modern mechanic.

Small as it is, the smallest possible fraction of a second, almost inconceivable to the human mind, there is a distinct difference between the time at which the spark passes across the ends of the wires in the spark plug, and the time at which the force of the explosion begins to take effect. To insure the greatest efficiency, the ex-

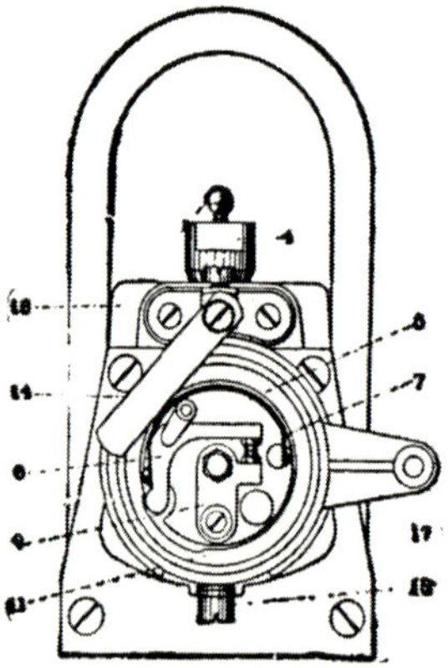

Contact Maker on Bosch Magneto.

plosion should occur at the exact second when the piston has reached the limit of its upward stroke and is about to begin its downward action. As there is an appreciable interval of time between the sparking and the ignition, or explosion, it follows that the make or break of contact which produces the spark must be made just that much in advance of the time when the piston is ready to receive the impact.

How the Spark is Regulated.

In the high-tension magneto this regulation of the spark is secured through the manipulation of the "timer," which is a part of the magneto, and controls the flow of the electric current, or in techincal parlance, the "impulse." It will aid the novice in a better realization of its workings to know that in producing this "impulse," the magneto acts directly opposite to the battery-and-coil system. In this latter the spark is made by the opening or breaking of the circuit. In the magneto the

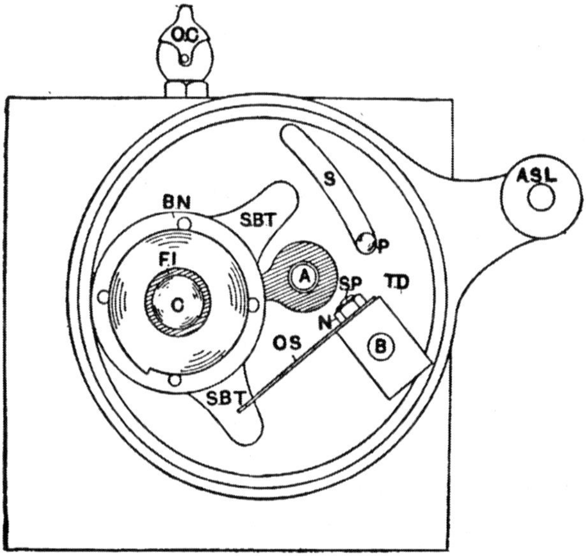

Timing Disc Without Contact Maker.

C plug in armature shaft; F 1 fibre insulation; S C steel cam; B N brass nut holding; S B T springy brass tongues; T D timing disc; F S forked steel piece (see fig. 44); P pin working in S slot; A and B steel pins; C S contact spring, held by N nut and S P split pin; O C oil cup; A S L advance spark lever.

spark comes when the circuit is "made" or closed. On one end of the magneto (see illustration) will be found a round brass box with a projecting lever, which in most motorcycles is connected by wire or steel rods to some convenient part of the machine, where the rider may advance (quicken) or retard (slow) the spark, while the machine is in motion. There are times, however, when this sort of mechanical control does not produce the desired results, and it becomes necessary to get at the mechanism of the timer itself. This is done by taking off the cover of the brass box referred to. This done, the mechanism is exposed and easy of access. It constitutes the contact maker and breaker, and is of exceedingly delicate construction, as will be seen by reference to the illustrations. By moving, for instance, the lock nut which holds the contact breaker, and changing the

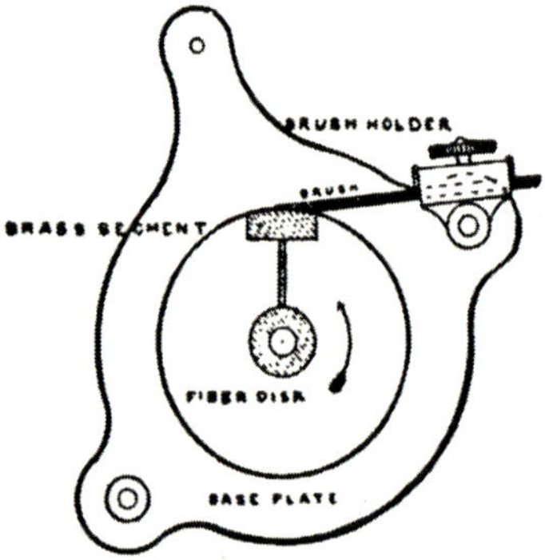

Another Form of Contact Making.

position of this latter the minutest fraction of an inch, a radical difference in sparking time is obtained. For this reason it is best to leave the manipulation of the timer to the attention of a competent electrician who has made a study of it, and has become a recognized expert in this particular line.

Battery Timing System.

When the battery-and-coil system of ignition is employed, the timing is done through the "contact maker." On the outside of the crank case, easy of access, will be found this contact maker enclosed in a case somewhat like the timer on a magneto. Open this case and you will see near the top a small strip of metal holding a platinum point, which is connected by wire with the spark coil. Directly under it is another platinum point similarly arranged and also connected by wire with the

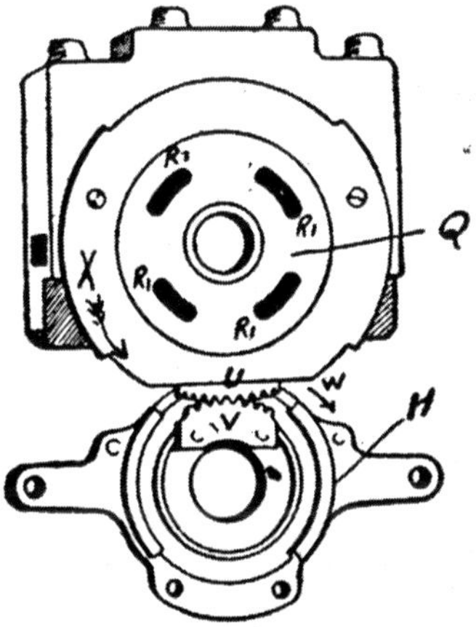

Timing Mechanism on Eisemann Magneto.

Q insulated distributing disc; R1 R1 R1 R1 distributing segments which when pressed, or touched by the revolving arm P (see front view of magneto) makes the contact (we are looking at rear of disc.)

battery. One of the gears in the crank case revolves a shaft, which runs through the center of the contact maker. On this shaft is a cam, or "eccentric" (see page 81). As the shaft revolves this cam exerts pressure of varying degree on the metal strip holding the lower platinum point. When this pressure is slightest, as is the case when the thinnest part of the cam bears against the strip, the latter naturally sags downward, separating the platinum points, when the current is interrupted, or "broken," because there is no contact. On the other hand, when the thickest part of the cam bears against the strip, it is forced upward until the platinum point it carries comes into contact with the one above it, and the current is given a clear field from battery to spark plug—an "impulse," as the electricians say.

Regulation of the contact maker is made by moving the position of the cam on its shaft, to which it is generally fastened with a set-screw, or some similar device. Note the mechanism of the contact maker carefully, as shown in the illustration, and its working will be plain.

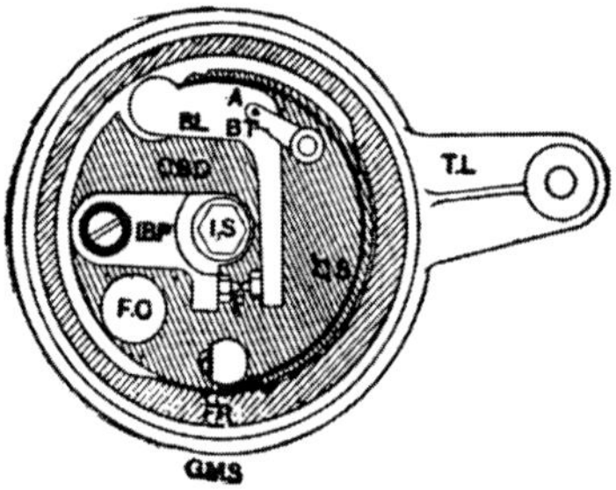

Timing Disc with Contact Maker.

C B D contact breaker disc; I S insulated screw; I B P insulated brass piece; P platinum points; B L bell crank lever; B T brass tongue; F R recessed fibre ring; G M S gunmetal strap; T L timing lever; C S curved spring; F C cover of frame connection from contact breaker disc.

Not only must the sparking action, whether obtained by battery or magneto, be properly regulated as to the exact time elapsing between each spark, but the engine valves must also be timed to act in unison with the sparks. There would be serious trouble, for instance, if the spark should come and an explosion occur when either the intake or exhaust valve, or both of them, were open. In the former case the motor would "back fire—that is, the exploding gas would be ejected backward through the intake valve with the danger of igniting the gasolene in the carbureter, and perhaps in the gasolene tank itself. On the other hand, if the exhaust valve were open, the force of the explosion would be lost. Consequently the valves must be timed to act in harmony with the sparking. This timing, however, is separate and distinct, and entirely different in its manip-

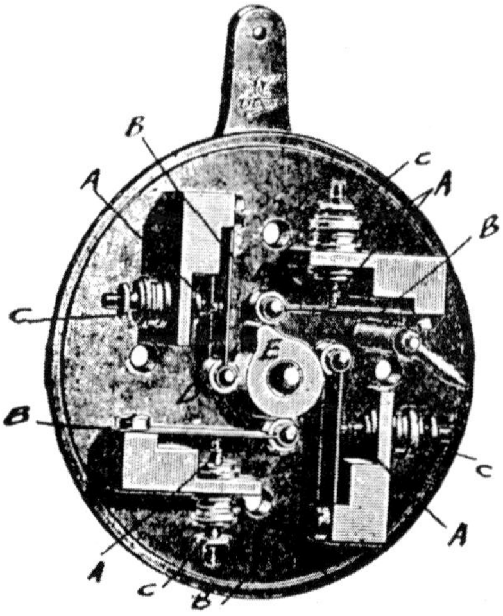

Still Another Contact Maker.

This form of contact maker is shown here merely because the parts are enlarged and their workings may be the more readily understood. The mechanism is as follows: A platinum pointed screws; B blades which are raised by the cam E, bringing the platinum points into contact; C lock nuts which hold the platinum pointed screws in place. If the spark comes too early the lock nuts are released and the contact screws placed a trifle further apart; if the spark is made late the platinum points are brought a little closer together. In either event when regulation is made the lock nuts must be set down hard so as to hold the contact points firmly in place.

ulation from the timing of the spark. It is, if anything, more involved than spark timing, and the man who would regulate the valve action of his machine must be something of a mechanic and mathematician combined.

The valves in some machines, and the exhaust valve in all, are operated by the action of gears contained in the crank case, which are set in motion by the descending and ascending strokes of the piston. Therefore, in order to regulate the movement of the valves, it is necessary to know the exact position of the piston. This is something which only an expert can do properly, and it is a good thing that occasion for resorting to it is of very infrequent occurrence.

Indian Valve-operating Mechanism.

Look at Figure 1 in Chapter VI and note the position of the exhaust valve D. It has been closed by the action of the lever E, which has been allowed to drop downward because the pressure exerted by the eccentric cam on the shaft F, has been taken away from it. This shaft is revolved by a gear connection from the crank case, and as it moves around the eccentric is continually raising and lowering the cam E, which opens and closes the exhaust valve. Changes in the movement of this gear connection, so as to affect the movement of the valve must, therefore, be made through the crank case. This is one of the features of timing the valves.

To Change Valve Action.

Assuming that regulation of this kind is absolutely necessary, one of the first things to do is to determine the exact position of the piston head. This can only be done by disconnecting the sparking wire and taking out the spark plug. This leaves an opening into the cylinder through which a stiff wire or rod is stuck, letting it go

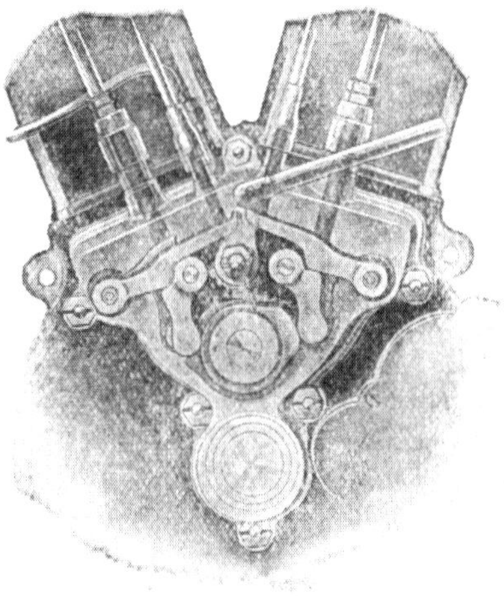

Mechanical Valves on Indian Twin Cylinder.

down until the end touches the piston head. Now move the fly-wheel or pedals slowly in the direction that will bring the piston upwards. So long as the rod continues to rise out of the spark plug hole it indicates that the piston is still moving upward. When it stops rising and remains stationary, it is an indication that the extreme limit of the upward stroke has been reached, and that the piston head is at the extreme top of the cylinder, minus the space left for the explosion chamber, which is generally about five-eighths of an inch. If the rod should begin to sink in the cylinder, turn the pedals or fly-wheel until the rod is again at a standstill at the top. Now make a mark on the rod even with the top of the outside of the cylinder. This done, use the pedals or fly-wheel again and lower the rod as far as it will go. Then make another mark on the rod even with the top of the cylinder. You will now have obtained the extreme range of the stroke—the upward and downward limits of piston action. What you are after is to determine the exact center. Having obtained the top and bottom extremes, it is easy to get at the center by measuring the distance between the two marks and dividing it by 2, allowing, of course, for the five-eighth inch dead space occupied by the explosion chamber.

You may next open the crank case—or do this before you begin taking the piston measurements. Remove the large gear which operates the valve cam. Next turn the fly-wheel so the piston is at center, which may be told by your rod scale. Then replace the large gear in the position shown in the illustration (note that the cam is just clearing the lifting lever). This results in the closing of the valve just as the piston reaches the end of its stroke, and secures the desired harmony of action. In other words, "perfect timing."

Fixed Timing in Vogue.

In most modern motorcycles, the timing of both spark and valves is fixed absolutely at the factory, and there is little chance for anything going wrong, unless in case of rare accident such as the breaking of some connecting part. Even men of experience who thoroughly understand the mechanism, favor this system. "What's the use," they argue, "of a man monkeying with an intricate part of his machine when it has been assembled and put in working condition by experts whose sole business it is to make it operate satisfactorily?" This is good theory, but there are times when it pays to know how to do it yourself; such, for instance, as in a case of emergency, when a bad fall, many miles from home, throws your sparking or valve mechanism out of gear, and the services of an expert motorcycle doctor are not to be had.

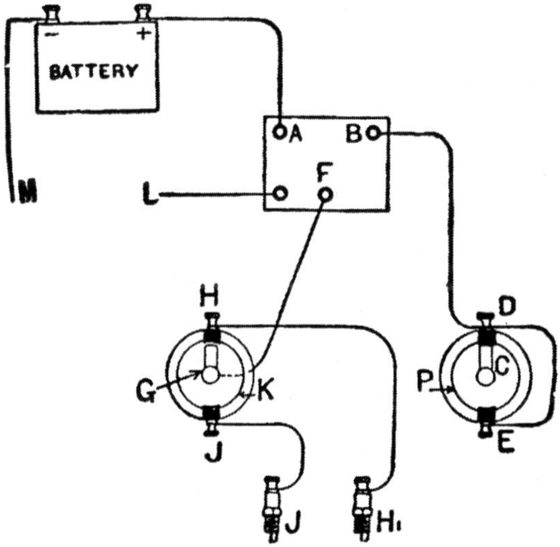

Wiring For Two-Cylinder Engine.

A, primary wire at coil; B, terminal; C, contact maker; D, E, places of contact; F, terminal with wire leading to insulated contact at G; H, J, connections with spark plugs, H1 J1; L, ground wire.

Occasionally the trouble with a valve may be located in the spring. This is sometimes too weak; if so, the valve will act sluggishly, and lose time. If the spring is too strong, the valve will act too quickly and thus gain time. A simple test may be made in this manner: Insert the tip of a screwdriver blade, or some similar tool, between the coils of the spring, and twist it slightly in the direction that will increase the tension. If, after this, the engine shows any sign, no matter how slight, of speeding up, it is a certainty that the spring is too weak. In testing for excessive strength in a spring insert the screwdriver blade on top of one of the coils and press downwards lightly, at the same time listening for signs of improvement in the engine movement. If these are noticeable, the spring is too strong. Before doing this, however, it will be well to make sure that the valve stem is not bent, and that it moves freely in its guide. If the stem is bent, or if it fits the guide too closely, the valve will be sluggish in action.

Timer on One-Cylinder Motor.

On one-cylinder motors the timer (it is not a commutator, although often erroneously referred to as such), consists of a cam which at the proper instant raises a spring so that it makes a contact with a stationary point, thus allowing the current to flow uninterrupted. When the cam pressure on the spring is released, the latter drops away from the stationary point and the current is interrupted, or "broken." Thus we get the name of "contact maker and breaker," which is really the proper designation for the device. As a matter of convenience, however, it is usually termed the contact maker. This device is generally mounted on the cam shaft, or half-time shaft, or on any part of the mechanism which is rotated at one-half the speed of the engine shaft. In this manner we get a contact every time the piston completes its four-cycle operation.

Just a Hint for Experts.

For the guidance of those who are already experts in valve mechanism and its theory, the following from

Sparks, published by the American Motor Co., may be of interest:

Roughly speaking, in a motorcycle engine, the exhaust valve should begin to open when the piston is at least one inch from the bottom center, and remain open until it has passed the top center and descended about ⅛ of an inch. Just at the top centre there is still a slight pressure which is relieved by allowing the valve to remain open a little longer. The inertia of the gases in the exhaust pipe also materially aids in reducing the

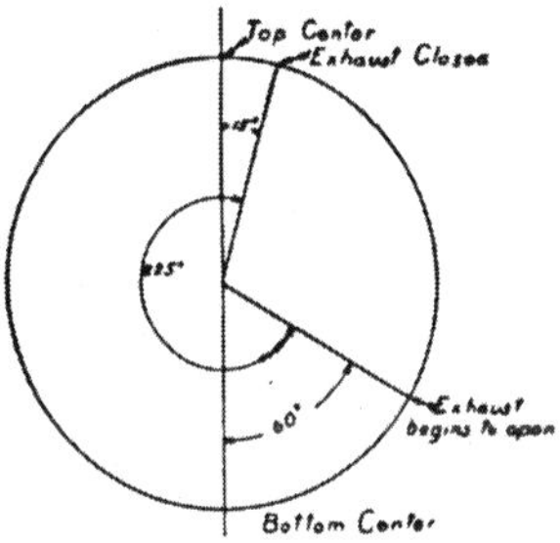

Fig. 1.

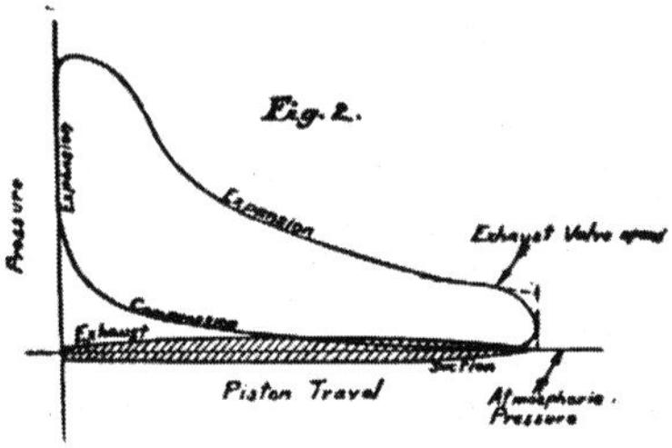

Fig. 2.

pressure or even causing a partial vacuum in the cylinder which increases the incoming charge.

The first diagram shows an exhaust valve diagram suitable for a motorcycle engine. It will be seen that the exhaust should remain open for about 300 degrees. We can best appreciate just why this setting is necessary if we examine the indicator card. This is a diagram showing the pressure in the cylinder at every part of the stroke. The pressure multiplied by the distance through which it acts is the work done.

In the second diagram the work done is shown by the white area inclosed in the curve. The lost work of charging and discharging the cylinder is shown by the shaded area.

This is a diagram taken from an engine with a valve setting like that of the first. It will be noted that the

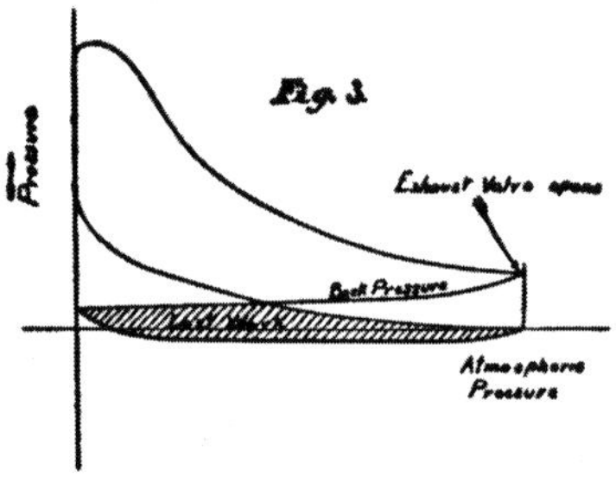

exhaust valve opens before the end of the stroke, and that the pressure at once begins to drop, thereby losing the small amount of work shown by the shaded area.

In the last diagram the exhaust does not close until the end of the stroke. The back pressure is thus much greater, and also the lost work. This is the reason why an improperly timed engine, or one with poorly designed cam, will not give the full power.

CHAPTER VIII.

THE CRANK CASE—ITS MECHANISM.

We have now traced the operation of the motorcycle from the vaporizing of the gasolene, by successive stages to the delivery of the power at the crank case, through the medium of the connecting rod. The question naturally arises—Why have a crank case? Why not let the crank shaft perform its duty direct without enclosing it? This would be practicable were it not for the fact that the crank case covers a multitude of important points.

Interior of Yale Crank Case.

Within it, by way of illustration, is contained the gears connected with the valves and magneto. It is one of the duties of the crank case to protect these gears from dirt and obstructions, to protect them from becoming entangled in the clothing of the rider, or from picking up

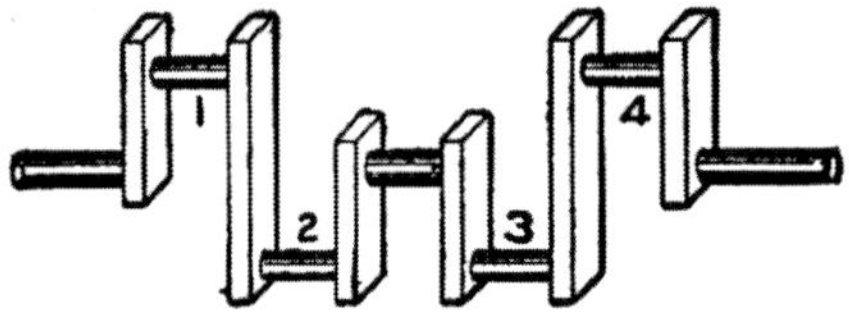

Crank Shaft For 4-Cylinder Motor.

trouble in the form of loose wires or other odds and ends along the road. The crank case on most machines further provides the receptacle for the oil with which the cylinder and other friction-producing parts are lubricated. Its bottom part forms a well which always contains, or should contain, a supply of oil at a certain level. In its

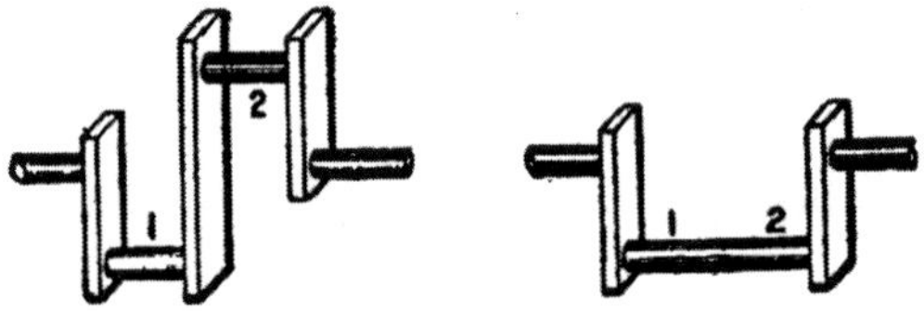

Two Ways of Coupling a 2-Cylinder Motor

movements the connecting rod—the lower end—dips into this oil and throws it up into the cylinder and other parts, thus producing what is called the "splash" system of lubrication.

Lay the crank case apart, and we also find that it covers the fly-wheel, which, in the case of the motorcycle, is made in two parts, the connecting rod being coupled to the crank shaft between the sectional halves. This fly-wheel is an essential part of motorcycle mechanism. By its operation any tendency of the piston to dead center is overcome, and a continuous motion is ensured, the wheel being heavy enough to "throw" the connecting rod over the center and bring the piston back into place for a new stroke.

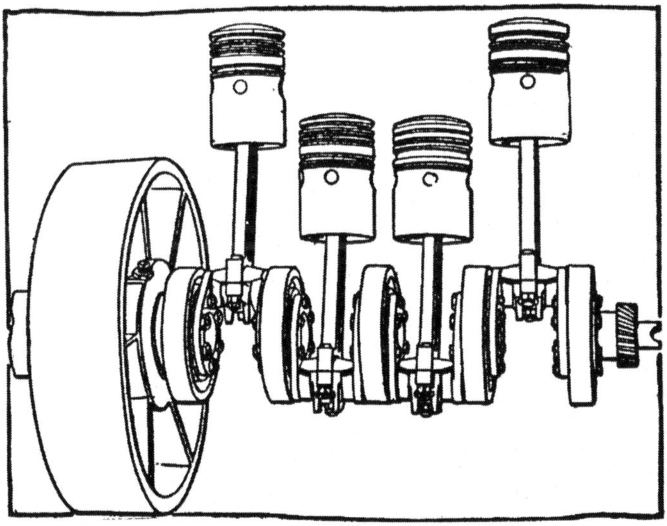

Four-Cylinder Crank Shaft Enlarged.

In some machines the arrangement is somewhat different. In the Harley-Davidson and Excelsior machines, for instance, a gravity system of lubrication is used, which does away with the necessity of carrying oil in the crank case, and the interior is consequently nearly dry. In the Harley-Davidson the crank case contains nothing but the crank shaft and fly-wheel, the gears which manipulate the contact breaker and valves being enclosed in an independent case attached to the outside of the crank case.

One and two-cylinder machines have the crank shaft placed transversely, or directly across the line of action. In the four-cylinder type—there is only one of these now on the market—the shaft is, of necessity, placed in a longitudinal position, as it is too long to place transversely. This makes the use of what is known as a shaft, or gear drive, imperative. The reader will find this part of the shaft-working further explained in the chapter on transmission.

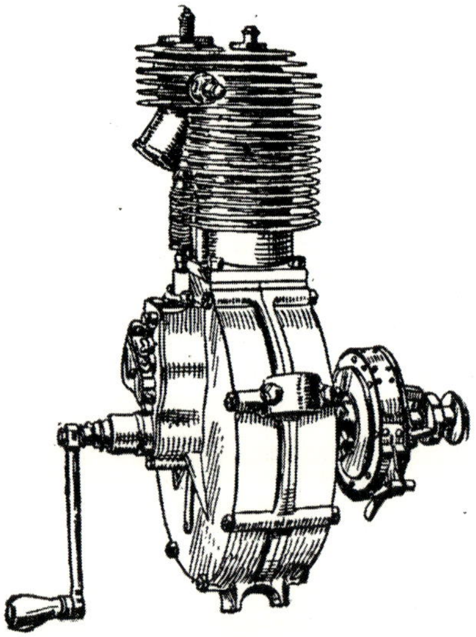

Hand Crank on Pedalless Machine.

CHAPTER IX.

TRANSMISSION OF POWER TO WHEELS.

Transmission is the term used to cover the connection of the engine power with the wheels. The "drive" is the method of applying this transmission. Riders and motorcycle makers speak of a shaft, or a chain, or a belt drive, This means that one of these forms is used to transmit the engine power to the wheels, in other words to "drive"

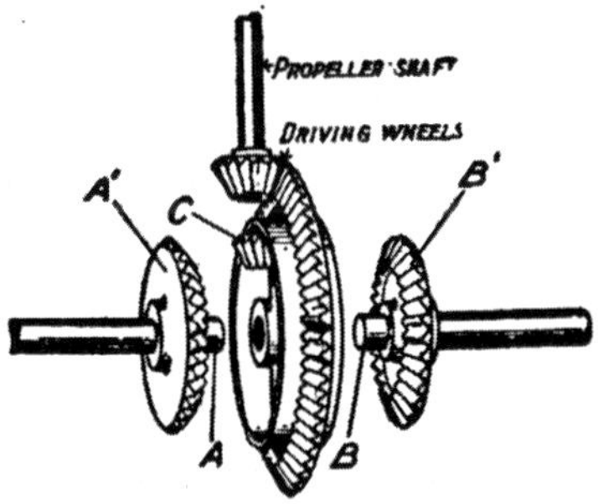

Pinions and Gears For Shaft Drive.

them. In motorcycles this transmission, as a matter of convenience, is to the rear wheel only. When this moves, under the impulse of the engine power, the front wheel simply has to get out of the way.

Shaft transmission at present is used almost exclusively on the 4-cylinder motorcycle, where, from the location of the cylinders in a longitudinal line, it is virtually a necessity as it would be difficult to use chains or belts. The shaft drive was used for a time on one and two-cylinder machines, but has been practically abandoned because of alleged lack of flexibility. In shaft transmission a straight steel shaft of the necessary length to reach from the point of transmission to the hub of the rear wheel is employed. Both ends of this shaft are fitted with

Belt Drive on Greyhound Machine.

bevel gears and pinions. Similar bevel gears are also attached to the crank shaft and the axle of the rear wheel, and into these the pinions or gears of the shaft are placed. As the gear on the crank shaft revolves the motion is communicated to the shaft by the bevel gear, and the shaft itself also revolves, in turn communicating the motion direct to the wheel.

Chain and Belt Drives.

In a chain drive the point of transmission is a sprocket —a toothed wheel—fitted to the crank shaft. On the rear wheel axle is a similiar sprocket, but of larger size (See Chapter XIII). Connecting the two sprockets is a chain composed of small, open links about 1½ inches in length, fastened together with metal bolts or pins which are readily removable. This division into small lengths

Samples of Chains For Drive.

gives the chain flexibility—the power to bend and accommodate itself to turning around a comparatively small circle, such as the initial transmission sprocket. This would be impossible in a rigid shaft or a non-joint chain. The openings in the links fit over the "teeth" of the sprockets and in this way a strong leverage, or holding power, is secured. As the initial sprocket is turned by the crank shaft, the chain, its links held by the sprocket teeth, turns with it, and a "drive" is thus given to the rear sprocket where the chain acts in exactly the same manner.

When the belt drive is employed the crank shaft is

fitted with a pulley over which the belt travels communicating the power or "drive" to a stationary or fixed pulley of larger size on the rear wheel. Two forms of belts are used, the flat and the V-shaped. When the flat form is used the surfaces of both pulleys are flat; when the V-shaped belt is employed the surfaces of the pulleys are grooved to accommodate it, and increase the holding power.

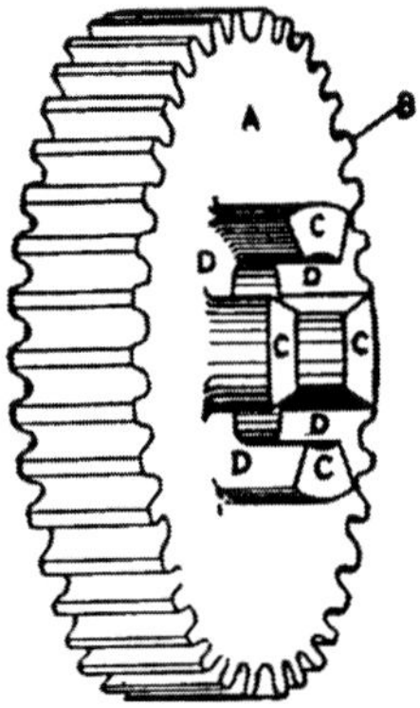

One Form of Clutch.

This is known as a "dog clutch." A gear wheel; B gear teeth; C C C C what is known as male clutch; D D D D female clutch. As the name implies the clutch engages or "clutches" a corresponding part of the mechanism. When thus "clutched" one speed is produced; when released the speed is changed.

Strong and Weak Points.

Each form of "drive" has its advocates; its strong and its weak points. Chains will break, and belts which are generally made of leather, will stretch and slip under certain conditions. In making a selection the novice should have in mind the character of the country in which most of his riding will be done. If it is hilly and rough, with roads full of stones and other obstructions, preference may well be given to the belt as there is less tendency to break under sudden jolts or severe strains. The belt also serves well in regions where there is no excess of moisture and the roads are reasonably dry. On swampy roads, or where the rain is continuous and heavy, leather belts will stretch and, although the slack may be overcome by the use of the idler pulley, or "tightener," it is a nuisance just the same.

In wet countries, for the reasons just stated, the chain will be found preferable. A link may break occasionally, or part at a joint, but this is readily and quickly repaired. On level, and reasonably smooth ground there is little chance for discrimination between the two sys-

Chain Drive Method of Transmission With Pedals.

tems. Both have flexibility, although the belt is unquestionably the more flexible of the two. This quality of flexibility is essential in the drive system of a motorcycle.

The man who is wedded to the chain system will say: "I don't have any trouble with my chain stretching or slipping."

To this the belt enthusiast will respond: "I don't have any trouble with my belt breaking under sudden strains."

V-Shaped vs. Flat Belts.

There is reason in both claims, and so well is this understood by manufacturers that most machines may now be had with either chain or belt drive as the fancy of the purchaser may dictate. Just now the problem is to make a choice between the flat and V-shaped belts. The latter are rapidly coming into favor as possessing a greater efficiency than the flat variety. This is accounted for on the ground that the V-shaped belt wedges itself into the V-shaped grooves in the pulleys and thus exerts a greater pulling power or drive than it is possible to get with a flat surface. Besides this the wedging process reduces the tendency to slip, and thus conserves engine power.

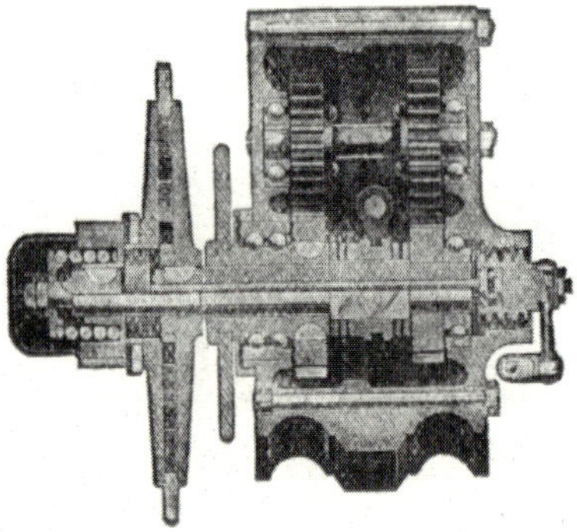

Two-speed Gear Device on Indian Motorcycle.

Two-Speed Transmission.

Some motorcycles are now being made with a two-speed transmission, an ingenious arrangement by which the machine may be changed from high to low, or from low to high speed, at the will of the driver without interfering with the action of the engine. This result is obtained by the use of a sliding gear device which is manipulated by foot levers, or by a lever handle within easy reach of the driver. In changing the speeds a clutch similar to that used on automobiles is employed. As the lever handle is pushed forward or backward, or the foot levers pressed, the clutch engages or disengages

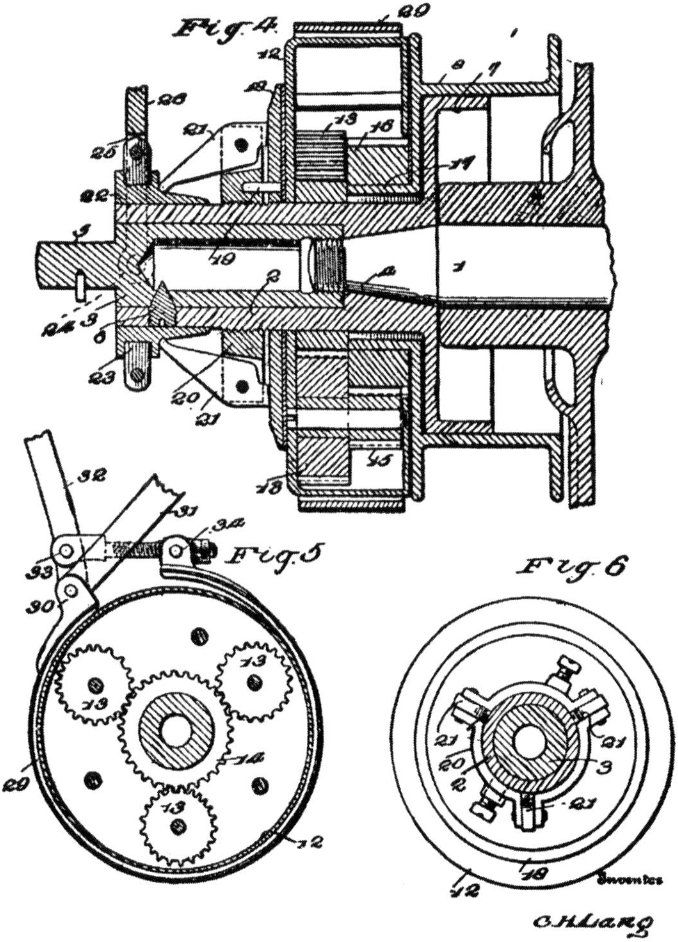

Side Elevation Lang Two-speed Transmission.

the speed gear. There are two of these gears, one of which produces a high, and the other a low speed, independent of the engine movement. The main advantage of this is in hill climbing, or the carrying of excessive loads where power is more of a requisite than speed. In instances of this kind the transmission may be shifted from high to low speed, leaving the engine

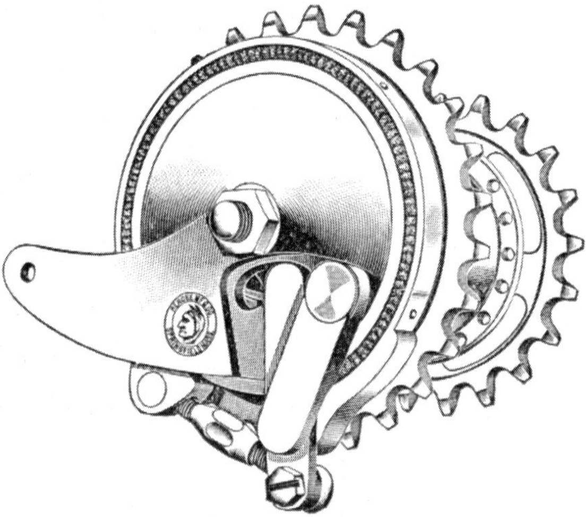

A Coaster Brake.

still working at top speed. In this way an increase of actual power is obtained, the engine energy going into the production of this power rather than that of speed. This two-speed transmission is also found handy when it becomes necessary to slow down, and yet keep moving, as is the case frequently in crowded streets, or when encountering some obstruction on a country road.

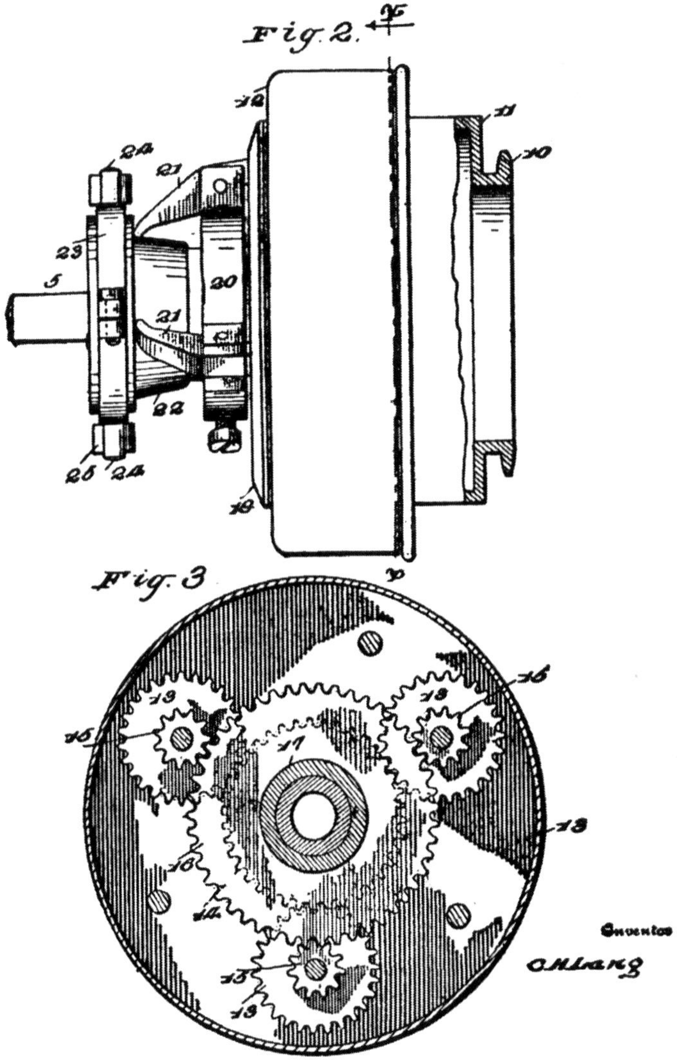

Longitudinal Section Lang Two-speed Transmission.

How Two-Speed Transmission Works.

The two-speed transmission used on the Harley-Davidson machine is the invention of Mr. C. H. Lang, a Chicago mechanic. In its operation the cone marked 22, (See Figure 2) is forced by lever atcion at the will of the order into the drive mechanism so as to lock the jaws 21, and thus force the friction disc 18 tightly against the face of the gear case 120. This engages the gears and produces the high speed.

In shifting to a low speed the clutch disc 18 is released by contrary lever action, freeing the high-speed gears and engaging those which produce the low speed.

The Pedal Drive.

While the "drive" on motorcycles is from a small to a large pulley by chain or belt as the rider may elect, nearly all machines are also fitted on the opposite side with a special chain drive connecting a sprocket on the crank shaft with one on the rear axle. This is for use in case of emergency, and is known as the "pedal drive." If an engine should "balk" on the road, and the rider be unable to get it working again he may disconnect the engine drive and pedal the machine home. It will be noticed that these "pedal drive" sprockets are of exactly opposite proportions, as regards size, from those used with the engine drive, the one on the crank shaft being larger than that at the rear axle. This is because the engine produces more speed than can be utilized with comfort or safety, and it is necessary to reduce it by making the rear drive pulley larger than the initial transmission pulley. In the case of foot power (pedalling), however, the rider is not able to produce speed enough, and consequently a reverse process is resorted to, the rear sprocket being made smaller. Thus the speed produced at the initial transmission sprocket is increased in ratio to the reduction in the size of the rear sprocket. If the initial sprocket is, say 8 inches in diameter, and that at the rear axle 4 inches, the latter will develop twice the rate of speed because it makes two revolutions to every one made by the former.

Where Pedals Are Eliminated.

In one machine, the New Era, the pedals are eliminated. It is to all practical purposes a two-wheeled auto-

New Era Machine—No Pedals.

mobile, the mechanism being strikingly similar. Compression for starting the engine is obtained by means of a crank attached to an end of the shaft projecting through the crank case. This machine, which is called an "autocycle," is also fitted with the two-speed clutch transmission, and has other features which hitherto have been used in automobiles only. When asked how, in the absence of pedals, a rider would get home in event of a serious breakdown, the Chicago representative of the maker replied:

"How would an automobilist, caught in a similar predicament, get home? Remember our machine is not in reality a motorcycle; it is an automobile on two wheels—an evolution of the motorcycle and automobile combined. It has this advantage, however, over an automobile. In case of emergency the auto-cycle may be picked up and loaded into a wagon. But why worry about this? Modern machines are not supposed to break down, so far as the engine is concerned, in a manner which precludes at least temporary repair."

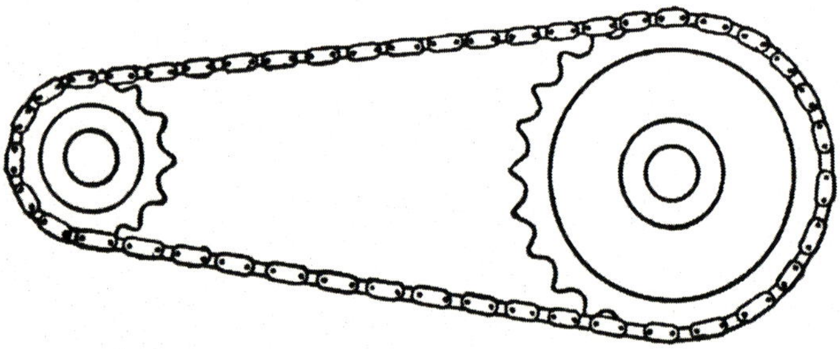

How a Chain Drive Looks.

CHAPTER X.

LUBRICATION—VARIOUS METHODS IN USE.

All machinery requires thorough lubrication, and none more so than that of the motorcycle. The points where metal parts move or revolve in contact must be kept well oiled. Lack of proper lubrication creates a destructive friction which not only wears out the various bearing parts, but interferes seriously with the efficiency of the machine. Friction produces heat, and heat makes metal expand, this expansion often resulting in a "choking" of the bearing parts. Take, for instance, a metal shaft revolving in a metal channel or bearing. These fit snugly so there is no rattle or lost motion. So long as these parts are kept well oiled the friction is reduced to the minimum. When the oiling is neglected and the parts run dry the friction increases and the intense heat thus generated swells or expands the shaft so it chokes the bearing and becomes stuck. All of us at some time have seen teamsters stalled by the sticking of a wagon wheel which would not turn. This is caused by an oilless axle becoming overheated, and choking the hub bearing. Exactly the same thing occurs in machinery which is not properly lubricated. The higher the speed the greater the friction, and consequently the greater the tendency of the bearing parts to swell and choke.

There is no machine, aside from the automobile, which is operated at so high a rate of engine speed as the motorcycle. The ordinary working maximum of the piston is 1,000 feet a minute, which means that in its upward and downward movements, made within a 3, or at the most, a 4-inch limit, the piston travels 1,000 feet every 60 seconds. The rapidly recurring explosions of gas in themselves tend to overheat the cylinder and, if this latter is not kept well lubricated the piston speed will create a friction which will greatly increase this heat. There are other parts such as the crank shaft, the gears, etc., to which this also applies with more or less force. All of these parts must be kept well oiled if trouble would be avoided. There is no necessity for "sousing" all of these parts equally. Some require a large quantity of oil, while others will work equally well with less. The cylinder and crankshaft need the most oil, and the supply should be liberal and unfailing, but even here, large as is the amount needed, there is danger of over-doing it. If there is a surplus of oil in the cylinder chamber, more than is actually required for proper lubrication, this surplus will be burned into carbon by the heat of the cylinder, and pre-ignition, and other troubles will follow. (See Chapter XII).

Quality of Oil Required.

First in importance comes the quality of the oil. This must have body enough—viscosity, the experts say—to afford the necessary lubrication under extreme pressure, and this pressure is extremely heavy in a motorcycle cylinder. If the oil is too thin the pressure will force it out from between the bearing surfaces—the piston head and the cylinder walls—and the lubrication will be insufficient. At the same time the oil should not be so thick as to have a tendency to "gum". Nearly every motorcycle maker has some favorite brand which he recommends as being particularly applicable to that one machine.

In selecting a lubricating oil the following qualities should be kept in mind:

It should be what is known as a high temperature oil —capable of withstanding an extremely high tempera-

Consideration should also be had for the method of lubrication used. In some machines the oil is forced directly to the bearing parts through small holes bored through the metal. In this case care must be taken that the oil is not so thick as to prevent its free passage through these channels.

How to Tell Good Oil.

Oil for lubrication should be pure mineral, without admixture of animal or vegetable substances. The heavier the oil (unless it is adulterated) the darker will be the color, and the darker the color, as a general thing, the greater will be the amount of carbon. Light-colored oil, that is oil that is naturally light and has not been bleached, is usually the best, provided it has the necessary fire and viscosity tests. Oil dealers understand the merits of a natural light-colored oil and some of them produce it to order by bleaching the dark, heavily carbon-laden oil with acid. When the presence of acid is suspected put a piece of blue litmus paper (to be had at drug stores) in the oil. If there is any acid present the litmus paper will turn a pinkish color. If litmus paper is not to be had saturate a piece of flannel or lampwick with the oil. Lay this on a polished steel plate and expose it to the sun for twenty-four hours. Wipe the metal dry. If the oil is free from acid the metal will retain its polish. If the oil contains acid, dull spots will show on the surface of the metal.

Reject oil that contains acid.

Making a Flash Test.

A flash test may be made by heating a small quantity of oil in a porcelain vessel over a Bunsen flame. Stir the oil as it is heating with a thermometer that will register at least 420 degrees Fahrenheit. Pass a lighted match

Viscosity—210 to 230 at 70 deg. F.
Flash Point—410 to 420 deg. F.
Cold Test—15 to 20 deg. below zero, F.

Methods of Applying Lubrication.

Two distinct systems of applying lubrication are in use. These are the "splash," and the "positive" methods. They not only differ radically in action, but are in turn sub-divided as to methods of application.

The "splash" system is the oldest, and the one in most general use. In the bottom of the crank case a supply of oil is carried. In its rapid movements the lower end of the connecting rod dips into this basin of oil and throws, or "splashes" it onto the crank shaft, the gears, and up into the cylinder. The word "splash" is really out of place in this connection as the motion of the rod is so great that the oil is thrown up in the form of a fine spray or mist which penetrates to every part of the motor.

It is in the manner of getting the oil into the crank case that variations in this system come into play. In nearly every instance the oil is carried in a tank attached to some part of the frame, but each individual maker has his own ideas as to the manner of "feed." Some depend entirely upon simple gravity, and let the oil run directly from the tank into the crank case through a small pipe. Others use a drip feed, a regulator on the feed pipe, which allows the oil to drip into the crank case at a fixed rate. When this is used it is generally accompanied with a small glass gauge, placed at the same level at which the oil in the crank case should be maintained. This is a convenience as by looking at the gauge the rider may always tell whether he is carrying the proper amount of oil in the crank case. Then there is still another class of manufacturers who employ what is known as a pump feed, holding that it makes the necessary supply of oil more certain and positive. This pump acts automatically, the power being taken from the motor.

Applying the Oil Direct.

In the "positive" system the oil is applied by an automatic pump direct through a small hole, to the crankshaft bearing. The crankshaft is hollow and the oil works its way through the shaft to the crank-pin bearing. It is next thrown by centrifugal action into the lower part of the cylinder. A circular pocket in the crank case gathers the oil and delivers it to the cylinder wall, piston and wrist pin. The oil pocket at the base of the cylinder is of such form that the piston at the completion of its downward stroke enters the pocket, picks up the oil therein and on the upward stroke spreads it evenly over the cylinder wall.

It is claimed for the forced system that a considerably thicker oil may be delivered direct to the point of application in cold weather than would be possible with a gravity feed, especially where it is to run through small holes as in the "positive" method.

Effect of Cold Weather.

Oil never actually freezes like water, but it does congeal, or thicken in cold weather. This should be borne in mind, when making purchases of lubricating material. For the reason stated you can use a heavier oil in summer than you can in winter. At the same time the difference is not so great as might be expected because the cylinder and even the crank case are always warm when the machine is in motion and this heat has a tendency to prevent congealment. The trouble arises when long stops are made and the motor cools off. Then, if it is zero weather, the oil will thicken.

Trouble With Dirty Oil.

After oil is used for a time it becomes dirty, and dirty oil is a nuisance as it not only gums the bearing surfaces, but aggravates carbon deposits and pre-ignition. This dirty oil finds its way back into the bottom of the crank case. This should be drawn off occasionally, and always before putting in a new supply. Most crank cases are fitted with a cock in the bottom, or lower part, for this purpose. Where the direct gravity method of supplying the oil is used you first close the feed cock. Then, having gotten rid of the dirty oil, if you have plenty of time, you may let a fresh supply work its way into the crank case. But, if the saving of time is an object, as it generally is, a charge of sufficient size to ensure the proper level, (so the end of the connecting rod will dip into it) must be injected into the case with a pump.

Over and Under Lubrication.

There is such a thing as over-lubrication, but it is not so frequent as under-lubrication, especially in new machines which require more oil than those which have been used for a time. Both are evils, but of the two under-lubrication is certainly the worst. Lack of oil, as already explained, wears out the motor and its bearing parts, and makes the buying of new parts or expensive repairs necessary. Over-lubrication causes carbon and pre-ignition, but these are readily overcome. Too much oil will manifest itself in a thick blue smoke issuing from the muffler. When this occurs disconnect the sparking wire and take out the spark plug. Pour some gasolene into the spark plug opening, and turn the pedals a few times. This will wash out the surplus oil through the exhaust valve.

Possible Use of Graphite.

There is a possibility that in time the necessity for exclusive oil lubrication may be overcome, so far as the cylinder itself is concerned. If this ever occurs it will be by the use of graphite. This substance, although a solid, has great lubricating properties. It has more resistance to pressure than oil, and is not so readily forced out of

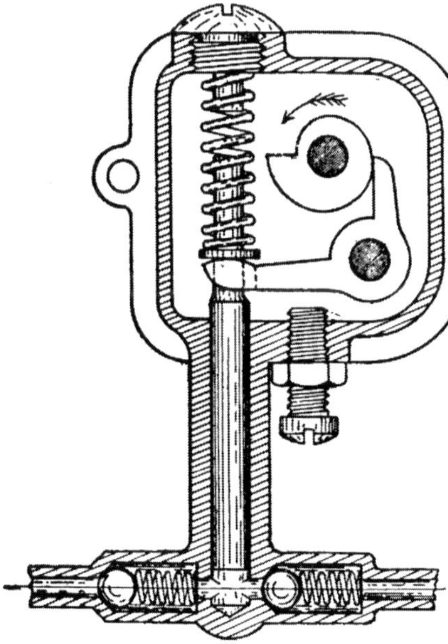

Merkel Mechanical Lubricator.

the cylinder. In fact, when once the cylinder walls are coated with it there is no normal pressure such as that existing between the piston head and the cylinder which will remove it. Graphite has the further property of filling in and making even and smooth such worn or pitted surfaces as may exist. The surface it presents is thin, but wonderfully smooth and velvety, reducing friction to the smallest possible amount. And yet in its use we run against a peculiar proposition. Graphite is plumbago or black lead ground to an impalpable powder. It is a pure carbon and as such would seem to be objectionable as tending to cause the troubles which originate from oil carbon. Despite this experienced men are using it with success, and there appears to be a possibility that this use will increase. It would not be well, however, for the amateur to experiment with it.

CHAPTER XI.

ROAD TROUBLES—HOW TO OVERCOME THEM.

No matter how perfect a machine may be, or how experienced the rider may become, there are times when troubles will arise. As these almost invariably occur when least expected it is well to know how to act in case of emergency. The rider should be somewhat like a physician, able to diagnose the symptoms of his patient, and qualified to apply the proper remedy. While some symptoms point directly to the nature of the trouble, there are others which may have their origin in any one of half a dozen causes, while at the same time one cause may manifest itself in half a dozen different ways. Here is where a thorough understanding of motorcycle mechanism and principle is of great value. Our purpose in this particular chapter is to describe the more frequent troubles, analyze them briefly, and show the man who is thus beset what to do.

Sudden Stoppage of Engine.

Generally this sudden engine stoppage is a case of "no spark." See if the switch is "on," and making good contact.

If you are sparking with a battery system, see that the amperage is sufficient to induce a suitable spark. Also see that the spark plug is properly set (screwed down) in the cylinder head. If you do not locate the trouble, look for loosened or broken connections at switch, battery and coil. Sometimes the insulation on the wires wears away, causing a short circuit.

Occasionally the rider will locate the trouble in the

contact maker or timer. This may arise from the lever becoming stuck, an insufficient contact, a broken spring, or shifting link disconnected.

In slightly varying forms, the same sources of trouble may be traced in the magneto sparking system, and overcome in the same manner.

There are other causes for the stoppage of an engine, or its refusal to start, which may be classified as follows:

1.—Broken or jammed gears.
2.—Gasolene accidentally shut off.
3.—Improper gas mixture.
4.—Intake valve stuck.
5.—Lack of compression.
6.—Water or dirt in gasolene.

It will be best to look over the parts first mentioned before taking up these latter. Having satisfied himself that the gasolene supply is running freely, and the ignition system in order, the rider should inspect the various features as numbered from 1 to 6. Nos. 2, 3, 4, 5, and 6 will be reasonably easy to overcome, once the cause of stoppage is located in any of them. No. 1 is a matter for the repair shop unless the rider has an extra set of gears.

Having located the trouble, the nature of the remedy, the means of overcoming the difficulty, naturally suggests itself. There should be no necessity, for instance, of telling a man what to do if he finds the insulation worn. Common sense will suggest a rewinding of the damaged part with the insulation tape, of which every sensible motorcyclist carries a supply.

Gradual Stoppage of Engine.

If your engine comes to a gradual stop, the first thing to do is to look into the gasolene tank and make sure you have a supply of fuel. In one well-authenticated instance a rider was stalled in this manner. He and his mates—it was a party trip—spent over an hour trying to locate the trouble before one of them thought of testing the fuel supply. He found the gasolene tank empty. A fresh supply was procured and the machine moved readily.

Shortage of fuel is the most prolific cause of gradual stoppage, but it is also occasionally found in a weakening of the battery, and consequent failure of ignition. Batteries are, in one respect, like a clock—they will run down in time. The longer a battery is used the weaker the amperage—or flow of the current becomes. There may be enough to move your machine properly when you start out, but it gradually fails until it is so low that the current is not sufficient to make the spark. When this occurs you must replace the exhausted cell; there may be only one, and on the other hand all three may be exhausted.

Engine Fires Irregularly.

Much of the irregular or intermittent firing (explosions) may be overcome by keeping the drain drip on the carbureter clean, as a great deal of the trouble is directly traceable to dirt obstructions. If there is a drain drip on the gasolene tank, this should also be kept clean. If these drips are clean, and the intermittent action continues, the rider may look for the seat of the trouble in some one of the following causes:

1.—Broken insulation on wires.
2.—Inadequate fuel mixture.
3.—Cracked spark plug.
4.—Defective electrical connection.
5.—Gasolene feed choked.
6.—Moisture or dirt on spark plug.

7.—Poor contact in timer.

8.—Improper action of valves.

9.—Terminals on coil loose or damaged.

As a general proposition, the trouble may be traced to a dirty spark plug, or defective connection with the ignition system. The rider should first detach the wire from the spark plug and remove the latter for inspection. If it is perfectly clean, and the wire points at the "gap" not too far apart, he may replace it and look for trouble elsewhere as indicated. If the plug is dirty, clean it before replacing. If cracked put in a new one. Should the engine continue to fire irregularly, it is plain the fault lies in another direction, and he should then go over each of the parts carefully, as given in the list numbered from 1 to 9, omitting Nos. 3 and 6, as these will have been covered by inspection of the spark plug. The trouble once located the manner of remedying it, as previously stated, will be plain. It should not be necessary, for instance, when the fuel mixture is found to be too weak, to tell a man that he must enrich it by adding more gasolene, or when there is a poor contact in the timer that it must be readjusted.

Poor contact in the contact maker, due to worn or pitted condition of the platinum points, will cause "skipping." These worn points will work all right for a time, and then suddenly fail to come together. When this happens there is no contact, and no spark.

When the Engine Knocks.

"Knocks," or "pounds" in the cylinder, while annoying, can hardly be classed as a road trouble as, while aggravating, they seldom interfere with the rider getting home. As a rule, they come from two causes—lost motion and pre-ignition, the latter doubtless being the more common. In either instance, the remedying of the "knock" is a matter for attention at home, or in the repair shop. As a temporary relief, when the cause of the pounding is pre-ignition, the spark may be retarded.

Where lost motion is the cause, it may generally be traced to an improper fitting, or excessive "play" between parts which should have a smooth, uniform motion. When this occurs, the parts must be overhauled and refitted.

Besides these, which are the most usual, there are other causes which produce "knocks" and pounding. These may be enumerated as loose fly-wheel, worn bearings,

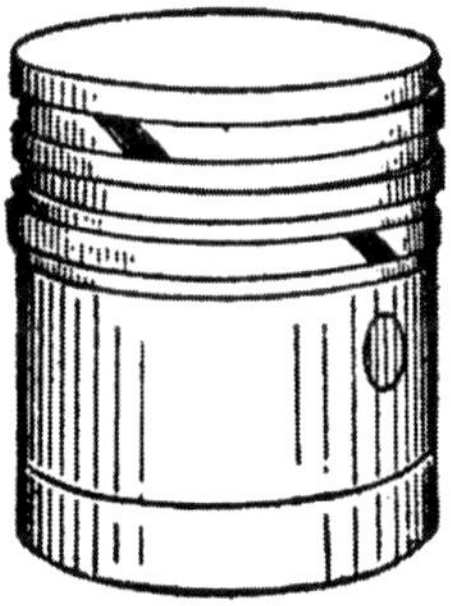

Piston Rings in Proper Position.

and occasionally a mixture that is too rich. The main thing is to know where to look for the source of trouble, and it may be found in some one of the causes mentioned. When an engine "knocks," no expert can say at a glance just where the trouble is, unless he happens to make a lucky guess. He must locate it definitely by looking for it, and the novice can do this just as well as the expert, if he knows where to look. When it comes down to a matter of making the repair or adjustment, the expert, of course, will have the advantage of experience, but even in this the novice is not badly handicapped as two or three practical object lessons will put him at ease and give him confidence. Let the man who is stalled on the road be compelled to adjust a timer once, and he will pick up a lot of valuable "know how."

Lack of Compression.

Proper compression is absolutely essential. You cannot start your motor without it. Suppose that on a tour you stop for a time by the wayside, and on getting ready to start again find your motor will not move. This is a clear case of lack of compression. First look for external causes. It is possible these may be found in the failure of the valves to lift properly, or in an overheated engine. If the valves work properly, next examine the intake valve cover to make sure there is no leak. The loss of compression power through leakages around the valve covers and seatings is surprisingly large, especially if they have had hard usage. In this case, the only actual permanent remedy is in re-grinding.

Worn piston rings are also responsible for the loss of compression. This rarely occurs to the extent of making it impossible to start the motor, but it does interfere with the speed and proper working of the engine. The best way of locating trouble of this kind is to run the engine at top speed for a short time. Then place your hand on the crank case. If it is unbearably hot the gas is leaking past the piston rings into the case. This may be due to the slots in the rings having worked into line. If so, it is easily remedied by rearranging their position. If the slots are found properly placed it is probable the rings are badly worn or broken and new ones will be necessary. Sometimes the ring slots will be found clogged with a gummy deposit from the burnt gas and oil, which acts as a cement, holding the ends together and preventing the expansion which is necessary to fill the cylinder space and make it airtight. Or the rings may have become rusted and immovable. In either of these conditions, take out the spark plug, bring the piston head up to the top of the cylinder and pour in a half a cup of kerosene. Let this stand in the cylinder for about a quarter of an hour. Having done this, work the pedals so as to bring the piston up and down rapidly. This will cut the gum and rust from the rings and leave them clean, although in especially aggravated cases it may be necessary to repeat the dose before the obstruction is entirely removed. When the piston rings are clean the

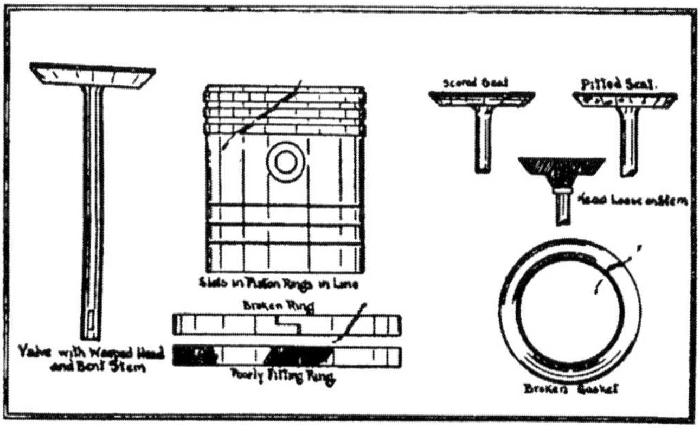

Common Causes of Lost Compression.

kerosene must be drawn off. Allow a few moments for the evaporation of what may be left, inject a little cylinder oil to insure lubrication and your motor will work all right.

Another cause of lack of compression is a leaky spark plug. If the plug is not screwed down snugly there will be a leak. Sometimes the joining of the spark plug to the cylinder is not tight enough, and nothing but the insertion of a gasket (packing) will correct the fault. If you suspect a leak at the spark plug, especially around the porcelain or mica, put on some oil or similar liquid. If there is a leak it will manifest itself in bubbles. When more screwing down, or the insertion of a gasket fails to stop the leak, put in a new plug.

Carbureter Troubles.

Aside from clogging with dirt you will have very little trouble with the carbureter itself. In case of stoppage, if the gasolene supply is ample and the ignition system in proper working order, it is reasonably safe to assume that the carbureter spray is clogged. No matter how much care may be taken particles of dirt will occasionally get into the gasolene and when they do they are liable, sooner or later, to work their way into the spray nozzle. The moment this is choked the gasolene feed is shut off, and the motor stops. In nearly every instance of this kind, the dirt may be removed in the following manner.

Take off the cover to the intake on the gasolene tank. Next open the spray nozzle as far as possible by means of the needle valve lever. Then open the air intake wide (or priming pin if there is one). This done, blow vigorously into the gasolene intake. This will usually force the gasolene through the spray nozzle with pressure enough to clear the small openings of the dirt which is clogging them. Regulate the needle valve lever to secure a normal feed, also the air intake, replace the cover on the gasolene intake, and pedal or crank to secure compression. In ninety-nine cases out of a hundred the motor will begin working again.

This process, however, it should be remembered, is only advised to secure temporary relief when you are on the road, and want to get to your destination. The dirt, as a rule, still remains in some part of the carbureter, and must be washed out.

Some wayside tinker may want to readjust the float valve, being sure it is out of order. Don't let him touch it. Don't touch it yourself until positive that it needs attention, which will be remarkably infrequent. It is a matter in which it will be real economy to consult an expert, a man who actually "knows." It is true that dirt occasionally gets into the float valve, but this is rarely the case. When it does occur, there is sure to be trouble

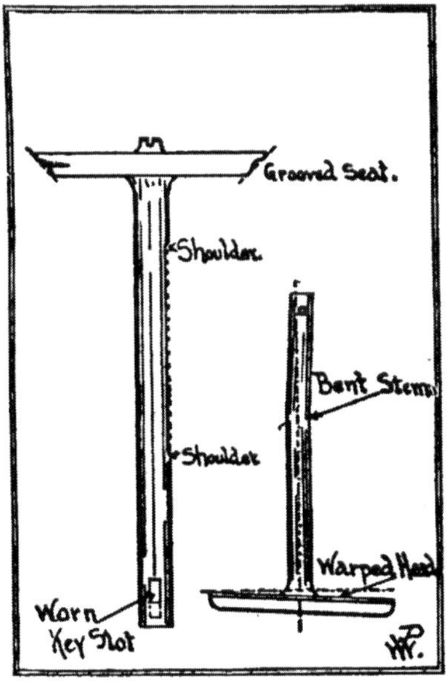

Worn and Bent Valve Stems.

as the valve will be held open and the gasolene will flood the carbureter. It is much more probable, however, that the obstruction will be found in the spray nozzle.

Popping Sound in the Carbureter.

At times the rider will hear a popping sound in the carbureter, but the cause is not in the carbureter itself.

It is directly traceable to the intake valve, the spring of which has become weak. This results in the valve acting too slowly, so that it is still partially opened when the explosion of a new charge of gas occurs. Part of this new charge is "back fired" into the carbureter, and the popping sound ensues. Strengthen the valve spring, or put on a new one of the proper tension, and the popping will cease.

Leaky Joint in Intake Pipe.

Occasionally, not often, there will be a leaky joint in the gas intake pipe. When this occurs too much air will get into the mixture, and the gas will not be strong enough to propel the motor properly. The worst of it is that such leaks are difficult to locate, because, as a rule, they are unexpected. The method in general use is to coat the joint with oil or soapsuds, or even saliva. Move the engine and watch the joint closely. If there is a leak, the substance with which you have coated it will be drawn inward, and you can note this readily. It is a case for a new union or new gaskets, but temporary makeshifts to stop the leak may be resorted to. One rider, unable to find anything else, coated a leaky joint plentifully with a paste made of clay puddled with water. Around this he bound strips of cloth torn from his pocket handkerchief. He waited patiently until the clay paste hardened, and succeeded in completing his journey.

Troubles With the Tires.

As a general thing the rider will have more trouble with the tires than with the mechanism of his machine. This does not indicate any defect or weakness in the tires themselves. So far as human ingenuity can devise they are practically perfect as to texture, material, make and application, but still accidents will happen. There will be punctures and blow-outs, and similar troubles, none of them serious, but all liable to cause annoying delay unless the rider krows what to do, and how to do it.

One of the most frequent annoyances comes from punctures. These, as a rule, are caused by running over sharp-pointed articles such as tacks, nails, jagged ends of wire, broken glass, and even rough particles of stone. So long as the puncture is in the outer casing only, there will be no leak, but if it penetrates the inner tube, the air is sure to leak out and the tire become deflated or

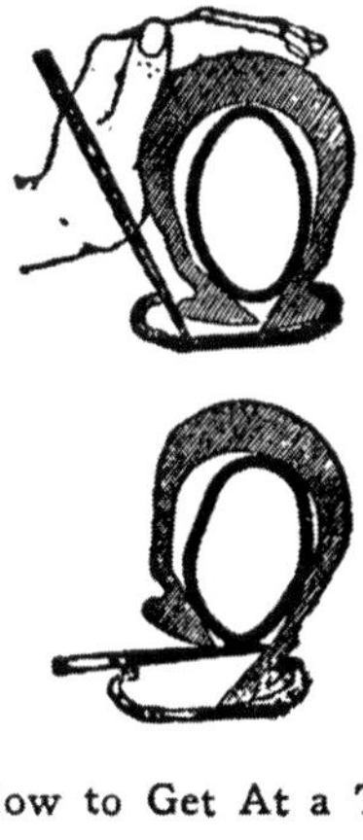

How to Get At a Tire.

"flat," so it is impossible to ride on it. This will be the better understood when we stop to consider the fact that all pneumatic (air-filled) tires consist of two separate parts—an inner tube of soft, thin rubber which holds the air, and an outer casing composed of a tough rubber mixture which takes the grinding wear of travel and protects the inner tube. Consequently, if this outer casing only is puctured slightly the rider may go on for a time without hesitation, but if the puncture reaches the inner tube, he must stop and make a temporary repair.

Repairs to Outer Casing.

This, also, holds true if the puncture or tear in the casing is of sufficient size to suggest the possibility of the inner tube, relieved of the necessary protection, "blowing out." Remember, that this inner tube is made of rubber, thin and delicate in texture, and not designed to withstand a heavy strain without support. When the outer casing is damaged in the manner stated, it requires immediate attention. Wise motorists will carry a supply of "repair plugs" and patches. If the hole in the outer casing is small enough, one of these plugs may be inserted and bound with tape.

On automobile casings protection is supplied by means of leather guards known as "sleeves," which are bound on over the casing and wheel rim. It is, as a rule, difficult to use these on motorcycles, because there is not clearance room enough between the tire and the frame. On an automobile the tires are outside the frame of the machine, but on a motorcycle they move between the forks of the frame reaching to the wheel hubs, with just clearance room to admit of the passage of the tire. An ordinary outer sleeve would "choke" this passage. At least one enterprising firm is meeting the demand for such a device by placing on the market a thin outer "sleeve" which is laced to the rim. It is a handy thing for securing temporary relief in a moment of pinching need. When the casing is opened, or torn in any way, the best, and about the only method of making a repair, is to put a canvas or thin leather patch on the outside, and a heavier material on the inside, directly under the rent, so as to protect the inner tube.

When Ingenuity Must Be Used.

Scheme and provide as he may, there will come occasions when even the experienced motorist will be caught without the necessary means of outer casing protection. Then he must rely upon his ingenuity. There was a time when the writer found himself thirty miles from home, on a country road, with a badly torn casing. He had no leather or canvas with him, and the prospect of making suitable headway seemed dubious. The tent of a camping party near the roadside suggested a temporary makeshift. For a quarter he bought a bit of the tent "flap," and with this, and the aid of a pot of cement which he carried, bound up the fracture so he got home.

Another rider in similar predicament, bound the frac-

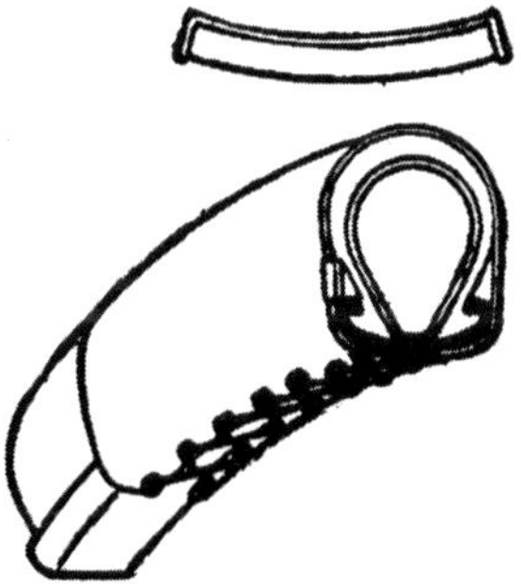

Sleeve Laced Ovei Tire.

ture with adhesive tape. But this needed protection and he was at a loss as to what to do until a bright thought came to him. He was wearing leather gloves with gauntlets. He cut off one of the gauntlets and used it as an inner protection between the casing and inner tube. It served admirably for a 12-mile ride. Still another rider of fertile resource utilized one of his linen cuffs. All of which tends to show that the man who gets along best is the one who uses his wits.

Here is an illustration of one rider who did not exercise this part of his mental makeup. He "picked up" a short nail which was firmly imbedded in the outer casing.

The head of the nail was even with the surface of the tread, and offered no obstruction to travel. There was no escape of air, and the rider could have gone along for miles without trouble. But he was evidently looking for it as he deliberately pried the nail out, leaving a fair-sized hole in the casing. Almost instantly the inner tube, relieved of support, blew out, and he had a long, tiresome job on his hands.

Bear in mind one thing: Every hole in the outer casing affects the inner tube and tends to weaken it and increase the danger of collapse. When such a hole or rent appears stop at once and close it in some way. If you don't do this, you will have more serious trouble.

Inner Tube Repairs.

All tires used on motorcycles are of the "clincher" pattern. This makes the inner tube comparatively easy to get at as the entire tire does not have to be removed. Raise wheel from the ground and at point of puncture push the casing in from the rim, having first partially deflated the tire which will loosen the pressure which holds it to the rim. Some flat instrument like a screwdriver or dull case knife, may then be inserted under the casing and used to pry it out over the rim. All casings are open the entire length on the bottom side which rests against the rim. Insert your hand in this opening and pull out inner tube. You are then in position to make the necessary repair.

In case the inner tube is punctured you will have to apply a patch, and over this place a protection of some kind. The surface of the tube should be first wiped clean with a rag, moistened with gasolene. When dry rub the spot lightly with sand paper to roughen the surface. This makes the patch stick better. If caught without a supply of the regular patches sold by dealers, a bit of canvas, or a piece cut from an old inner tube may be used. In any event you must have a pot of rubber cement. Trim the patch into circular shape so as to lessen the tendency to peel off which is strongest in a patch with square corners. Now smear the surface of the inner tube with cement at the spot where you are to apply the patch, at the same time coating one side of the patch. Allow these coatings of cement to dry thoroughly and then apply a second coat to both tube and patch. When this second coat has reached the sticky stage put the patch on, pressing it snugly into place with the fingers

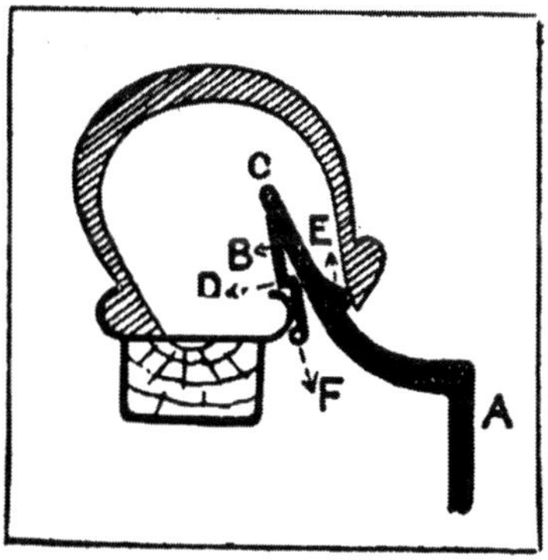

One Way to Pry Off a Tire.

A, specially made lever with strut piece; B, F, and shoulders D and E which bear against the rim of the wheel and the tire; C, hinge.

and then tap it well with a block of wood. Wait until the cement has "set" well and the patch is firmly in place before placing the protection over it. This latter may be almost anything that comes handy, a bit off your glove gauntlet or even a piece of thick cloth. Dust the tube well with French chalk, push it back into the casing, replace the latter on the rim, and re-inflate the tire. In case of emergency the tube may be wound with tape or canvas before being replaced, but this is not necessary if you can obtain a bit of cloth, canvas or leather.

Too much stress can not be laid upon the fact that all roadside tire repairs are merely temporary makeshifts—there is no real permanent relief except in having the damaged part vulcanized. No rubber cement that will withstand any considerable amount of heat has as yet been produced. Friction heats the tires, especially if the

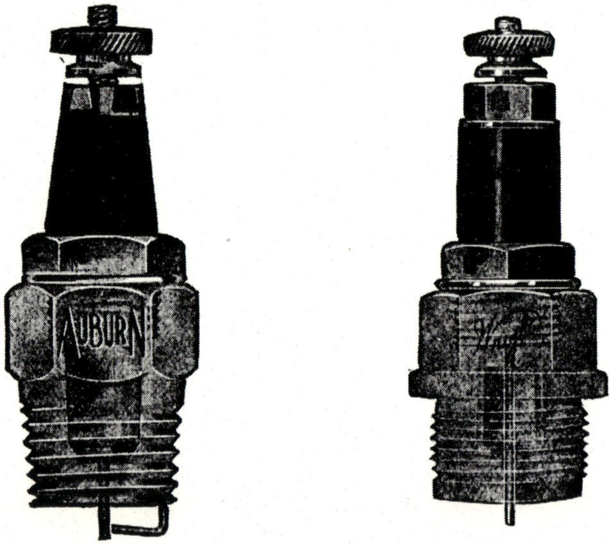

Two Forms of Auburn Spark Plugs.

pace is fast and the ground rough, and this heat softens the cement so that the patch works loose. Unless the rider realizes this he is likely to overlook the patched spot when he gets home and, because it served him well in a dilemma, start out on another trip with it. If he does, it is practically a certainty that he will get into trouble from the peeling off of the patch.

False Alarm of Puncture.

When a tire deflates without the customary sudden hiss, or the blow-out noise, don't be in a hurry to get at the tire. The trouble is much more likely to be located in the tire valve, whether the deflation is gradual or sudden. Before disturbing the tire, pump it up again and watch the valve closely. If the air is leaking through or around it there will be no difficulty in locating it. Frequently the nut which holds the valve to the air tube jars loose, or there may be a slight contraction of the rubber between the nut and the valve seat.

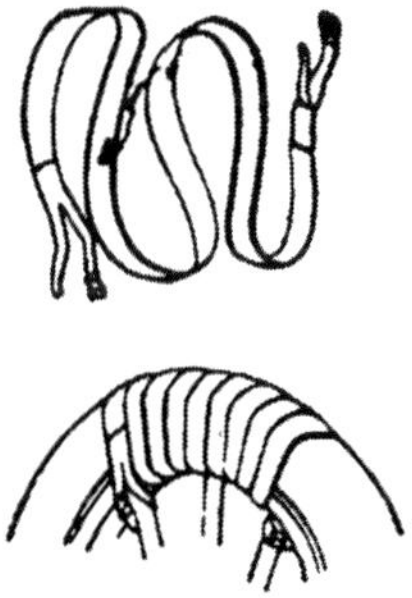

Repair of Tire With Strap.

Stopping Leaks in Tank.

Should a tumble (or other cause) result in a puncture of the gasolene tank put a tire patch over the hole, and tie, or otherwise fasten it with a bandage. Unless you do this the weight of the gasolene will force the patch off. Besides, the gasolene will dissolve the cement and the patch will drop off unless protected in some manner. It is at best only a temporary makeshift, and will last only a few miles, but it will serve in a case of emergency. If the tank seams leak they may be closed for the time being with chewing gum or soap.

CHAPTER XII.

PRE-IGNITION—ITS CAUSE AND EFFECT.

In pre-ignition, or the premature firing of the gas charge in the explosion chamber, the motorcyclist encounters one of his most frequent and annoying engine troubles. This advanced ignition results in forcing an expansion of the gas before the piston head is ready to receive it. In this connection, it is well to understand that it is not really the explosion itself which causes the piston to move, although it is customary to speak of it in this way. What actually occurs is this: In the explosion of the gas an expansion takes place. This expansion overcrowds the narrow space in the cylinder chamber, and something must give way. This "something" is the piston head which, being readily movable, responds to the expansive force produced by the explosion. If there was no expansion we would have a noise, but no force; the explosion would not develop any power, and pre-ignition would not be annoying. The cylinder mechanism is constructed to meet this expansive force at certain intervals and under fixed conditions, especially as regards the position of the valves. If the explosions occur too frequently and out of time (this is pre-ignition) there is a serious, unexpected strain brought to bear upon the connecting rod and its bearings; pounding and knocking ensue; there is a dangerous "back fire," and all the mechanism of the machine is in a way affected.

Three Causes of Pre-Ignition.

There are at least three well-known causes for pre-ignition. One is an ill-adjusted timing system, the spark being made too frequently (advanced too much) and the explosions being out of unison with the valve action. When this condition exists it may be overcome by retarding (slowing up) the spark so as to bring it into harmony with the valves.

Another cause of pre-ignition is projections or obstructions of some kind on the inside of the cylinder. In the case of a new machine it is probable these obstructions were not thoroughly removed in the grinding process. The rapid movement of the piston head against them generates a friction which causes the projections to become hot enough to explode the gas before the spark plug acts. In a machine which has been used for sometime without pre-ignition trouble, the obstructions may come from uneven wear in the cylinder, accumulations of dirt or rust, etc. The only satisfactory remedy is to have the cylinder reground, thus securing a perfectly smooth interior.

An over-heated cylinder will also cause pre-ignition.

The One Great Trouble.

The one great cause of pre-ignition, however, is dirt, under which heading may be included carbon. This carbon comes from imperfect combustion of the gasolene vapor, and the over-heating or cooking of the oil with which the cylinder is lubricated. If the gasolene mixture is too rich, if there is an excess of gasolene, the combustion will not be perfect. If there is an excess of lubricating oil, or if the oil is not of a quality to withstand a high temperature, the heat of the cylinder will cook, or burn it. In either event the result is a deposit of carbon. Much of this deposit may be avoided by using a gasolene mixture of proper proportion, and selecting a good quality of oil, the kind best adapted to cylinders at a high temperature. Delayed opening of either intake or exhaust valves also tends to produce carbon by cutting

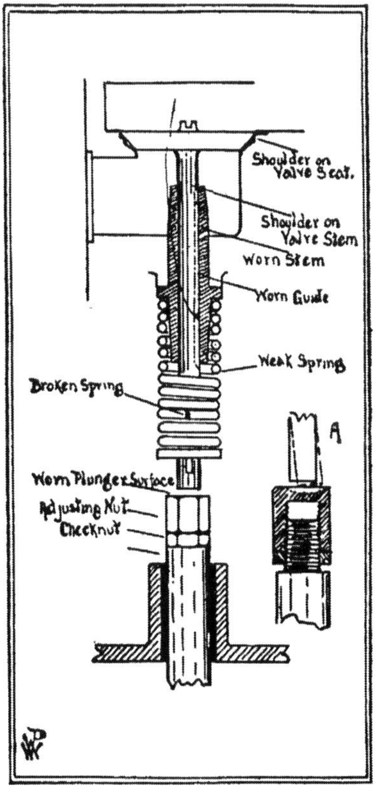

Valve Defects Which Cause Skipping.

At A we have an illustration of how a valve stem may "wabble" if the nut on top of the adjustable plunger wears hollow.

down the time necessary for complete combustion. So far as known it is impossible to avoid carbonization entirely.

The carbon, at first a gummy substance, is deposited on the piston head, and on the inside of the cylinder where it cakes and hardens. It often assumes the form of little lumps or points which become red-hot (from the friction) and fire the gas charges independent of the spark. Even when not heated direct by friction there is sometimes a tendency to spontaneous combustion. In either case, the result is pre-ignition.

How to Get Rid of Carbon.

Where this pre-ignition is due to carbon trouble, as is almost always the case, there is only one way to stop the nuisance, and that is to clean the cylinder and get rid of the carbon. This may be done the easiest and most effectually in mild cases in the following manner:

Take out the sparking plug, and pour a cup or half a cup of kerosene into the cylinder, having first brought the piston head up to the limit of its upward stroke. Then let the kerosene stand until it has had a chance to cut into and loosen the dirt. This, in some instances, will not require more than fifteen to thirty minutes. This "loosening" process will be hastened by working the pedals or turning the fly-wheel so as to move the piston head up and down quite rapidly, in this way securing an agitation of the kerosene which tends to make it cut into the deposit quicker. The best way is to let the kerosene stand in the cylinder over night, but the process first referred to will answer when the saving of time is a consideration, and the deposit is not too hard. Draw off the kerosene, or eject it through the exhaust. If it is very black and dirty it will be advisable to repeat the operation until it comes out reasonably clear. This done, let the cylinder dry thoroughly and then lubricate it with good oil, and the machine is ready to start again.

In unusually aggravated cases where the deposits have

become thoroughly caked and hardened, it will be necessary to take the cylinder apart and have the deposits scraped or chipped off. The piston head and cylinder itself should then be well soaked in kerosene for some hours when they may be readily scoured clean.

What One Man Does.

One genius, according to Motor Age, has evolved an ingenious scheme of his own for getting rid of the objectionable carbon deposits without resorting to the work of washing the cylinder. When this genius strikes a steep hill he opens switch and throttle wide and "coasts." The effect of this is to draw full charges of carbureted air through the cylinder and expel all unburned or partly burned gases from the muffler. This tends to cool the engine and at the same time has a strong cleansing action on the motor and exhaust passages, loosening and expelling the partly burned oil which produces the carbon. Cases have been known in which engines with a marked tendency to pre-ignition were handled in this manner and then ran for months without pre-ignition troubles.

Another Ignition Trouble.

There is another ignition trouble almost as annoying as pre-ignition, and this is mis-firing, which leads to an irregular, or "skipping" action of the motor. As a general thing, this may be classified as a roadside trouble, something which develops suddenly while the machine is in operation, and may be remedied temporarily at least, by some little repair made by the rider. Consequently it is treated of in Chapter XI on "Road Troubles," where the most handy means of obtaining temporary relief are described. Occasionally, however, there are cases in which more drastic or thorough overhauling is necessary, such, for instance, as cracked insulating bushings (the seatings) at the igniter. Still another fruitful cause of "skipping" is a broken valve spring. Even when a spring is broken there may be at times sufficient pressure to keep the valve to its seat; at other times the spring will not act as there is no tension. If this broken spring is on an exhaust valve, "skipping" is sure to follow as the valve will lift from its seat during the suction stroke when it should be firmly seated or closed. If, on the other hand, the broken spring is on the intake valve this latter, owing to the same uncertainty of spring action, may remain open when it should be closed. In either event "skipping," or irregular action will follow. The only real remedy is in the substitution of a new spring.

Sometimes when valve plungers of the adjustable types are used the top of the hardened nut may wear, becoming hollowed out where it comes into contact with the end of the valve stem. Such a condition, greatly exaggerated, is shown at A, in the accompanying illustration, the dotted lines indicating what may result if there is considerable side play in either the plunger guide or the valve stem guide. The valve stem may cramp over and hit the side of the hollow, and if the clearance has been lately adjusted by raising the hardened nut, the valve will be held open at times. This condition will cause irregular action under most circumstances, even if the other groups are in perfect condition. A bent valve stem, or carbon or scale on the stem, may cause binding, which in turn will cause skipping.

The illustration referred to is on page 137.

CHAPTER XIII.

HORSE POWER AND ELECTRIC POWER.

Engine efficiency, the value of its force, is universally expressed in varying degrees of horse power. Thus we refer to a 3, or 4, or 7-horse power motorcycle. This is popularly supposed to mean that the motor will produce power equivalent to that which 3, or 4, or 7 horses are capable of producing. But who can tell what a horse power really is? At school the pupil is taught that a horse power is the energy or force required to lift 33,000 pounds one foot in a minute. It takes only a few minutes of careful thought to realize that this is wrong. In the first place 33,000 pounds is 16½ tons. It would take an unusually strong horse to lift that amount of dead weight, even for one foot, to say nothing of sustained effort. James Watt was the originator of the term. What he defined as a horse power was the energy or force necessary to lift 550 pounds in a second. This is well within the actual power of a horse, but to continue the calculation indefinitely is ridiculous. The lifting of 550 pounds in a second most certainly indicates a power of lifting 33,000 pounds at the rate of one foot in a minute, but to state that the entire amount can be lifted at once by a horse is misleading. And yet this is the impression given to young people under the tutelage of those who are supposed to be careful, conscientious instructors. There is always a limit to the energy or force of a horse, or any other living creature, a time when it can exert its power no longer, but an engine, so long as it is in good working order, and has fuel, will run on indefinitely.

Relation of Power to Speed.

While power is an essential in the production of speed, it does not follow that the greater the power employed, the greater will be the speed attained. It is frequently the case that a machine of 3 or 3½ horse power will cover more ground in a given time than one of 4, or even 5 horse power. This result is governed by what is called the ratio of the drive. You must have power to climb hills, and the greater this power the easier the machine will get over the grade. Power is also a requisite in the hauling of a load—the more power you have the heavier the load you can carry. But this does not apply in the same way in the relation of power to speed.

In motorcycles the crank shaft carries a pulley or sprocket wheel fastened to the shaft. If a belt is used a pulley is employed; if the drive is by chain, it is a sprocket wheel. If the crank shaft is making 1,000 revolutions a minute it follows that the pulley will revolve at exactly the same speed. From this pulley a belt communicates the power to the drive wheel, securely bolted or otherwise fastened to the rear wheel. Now, if the pulley and the drive wheel were of exactly the same size, it is plain that they would both revolve at the same rate, and, as the rear wheel and the drive wheel are fastened together, they naturally move in unison, thus giving the rear wheel precisely the same number of revolutions.

Here's an illustration: The circumference of the average rear wheel of a motorcycle is approximately 84 inches, or 7 feet. If this wheel were propelled at exactly the same rate as the crank shaft pulley (at 1,000 revolutions a minute) moves—as would be the case if both pulley and drive wheel were of the same size—it would travel at the rate of 7,000 feet a minute, 420,000 feet, or 80 miles an hour. This would be too fast for comfort or safety. Consequently the speed is "geared down," as the experts say.

How Speed Is Regulated.

It is a well understood principle in mechanics that the transference of power from a small pulley to a large one reduces the speed quantity at the large pulley in exact

This Shows How Speed Is Reduced.

proportion to the difference in size. In motorcycles made for ordinary use (not racing purposes) the belt pulley or chain sprocket is about 16½ inches in circumference, and the drive wheel about 63 inches. This reduces the speed at which the rear wheel travels, because, while the transmission pulley is covering 1,375 feet in a minute, at 1,000 revolutions, the rear wheel, with a circumference of

84 inches, is only moving at the gait which the drive wheel (which is, of course, immovable) would make if **it took the place of the rear wheel, and actually revolved** on the ground. This would be 63,000 inches, 5,250 feet, or a trifle less than a mile a minute. Here we have a maximum speed of about 60 miles an hour.

When more speed is required, say, for racing purposes, the size of the pulley on the crank shaft is increased. The additional speed could be obtained equally as well by decreasing the size of the drive wheel, but this would be inconvenient, because, as already explained, this latter is a fixture, while the crank shaft pulley may be readily removed. In cases where there is too much speed (this seldom occurs) it may be reduced by using a smaller crank shaft pulley. As a general proposition, the ratio of increase in the size from the pulley to the drive wheel in road motorcycles is as 1 to 4, that is, the drive is four times as large as the pulley. In racing machines it is as 1 to 3, and sometimes even less.

One Peculiar Feature.

There is one peculiar feature, not readily understood, even by experts. A 6-inch crank shaft pulley, with an 18-inch drive on the rear wheel will give more power, but not so much speed as an 8-inch pulley and a 24-inch drive, both working with a crank shaft action of 1,000 revolutions per minute. The ratio is exactly the same in both instances—1 to 3, and yet the results are different. At first, this looks like a mathematical absurdity, but it exists, as every racing man and motorcycle expert knows.

Recognizing the necessity for having a more accurate and more easy system of calculating horse power, the American League of Automobile Manufacturers, embracing the cleverest mechanics in the country, has devised a plan based on piston displacement, or cylinder area. Under these specifications, 7 cubic inches of cylinder displacement constitute one horse power.

How to Figure Horse Power.

"But how shall I determine this piston displacement?" is the first thing the novice will ask. One way of doing this is to apply the following formula, which is supplied by Mr. C. H. Lang, a well known Chicago motorcycle expert:

This Gear Produces More Speed.
Note that pulleys are nearly of same size.

Multiply the cylinder bore by itself—if the bore is 3 inches multiply it by 3, which will give you 9. Multiply this product by 0.7854, and then again multiply this last product by the length of the stroke, the dimensions being taken in inches. Having thus obtained the total number of inches, it is only necessary to divide by 7, the number of inches required for one horse power, and you get the total amount of the latter. Thus, if the product of the combined multiplication shows a total amount of 21 inches, division by 7 will give the horse power as three.

Here is the way it works out: Say the bore of the

cylinder is 3 inches, and the stroke 3 inches; this is a fair average:

Multiply the bore by itself, viz:

3 multiplied by 3 equals 9 inches

Multiply the product by....................0.7854

7.0686

Multiply this by the stroke................ 3

21.2058

This is a total of 21.2058 inches of displacement. Divide this by 7, the number of inches of displacement required for one horse power, and we get:

21.2058 divided by 7 equals 3.0294. The result is a fraction over three horse power.

Another Method of Computation.

There is still another method of figuring horse power. This is: Take the square of the bore, which in case of a 3-inch machine would be 9 inches. Multiply this by the stroke which is, say 3½ inches, and we get 31.5. Multiply this in turn by 1800, and divide by 18,000, which gives a horse power of 3.15. Here is an illustration:

The square of the 3-inch bore, 3×3, is............9 in.

Multiply this by the stroke, 3½.................3½

31.5

Multiply by 1,800 1800

56.700

divide this by 18,000:

18,000] 56.700 [3.15

54.000

2.700

1.800—900

A Still Easier Method.

Those who desire an easier, quicker method of ascertaining the horse power of a machine, will find the following table adequate for all practical purposes. It is figured on the standard stroke:

Size of Bore.	Horse Power. 1-Cyl.	Horse Power. 2-Cyl.	Horse Power. 4-Cyl.
2½	2½	5	10
2¾	3	6	12
3	3 3-5	7 1-5	14 2-5
3¼	4¼	8½	16 9-10
3½	4 9-10	9 4-5	19 3-5
4	6 2-5	12 4-5	23 3-5

Engine speed, the rapidity with which the piston moves and the crank shaft revolves, has a direct influence on the exact amount of horse power developed. Hence the formulae in general use is based on 1,000 revolutions of the crank shaft per minute, which is about the maximum for a motorcycle. If this speed is reduced to 750 revolutions a minute, only three-quarters as much power will be developed as when a speed of 1,000 revolutions is maintained.

Computing Electrical Power.

There are a number of electrical terms which should be understood by the novice. Those which he will encounter most frequently are "amperes," "volts," "ohms," and "watts." These all have a distinct and reasonable meaning, and their use, while likely to be a little confusing at first, could not well be abolished.

Ampere is the unit of rate of flow or strength of the current, just as inch is the unit of measurement. It may be measured in various ways, but the usual method is by means of a specially designed meter.

The amount of "amperage"—the total of amperes required for the proper operation of the ignition system, is from 36 to 45. When the 3-cell battery is used this amperage should average from 12 to 15 amperes to the cell. The longer the batteries are used the weaker the amperage becomes. When it gets down to 4 amperes to the cell it can not be depended upon.

Volt is the unit of force or pressure. It may be best understood by citing the old example of water passing through a pipe. The gallon, for instance, is taken as the unit of measurement of the quantity, the same as the ampere represents the quantity of the electric current. When we refer to the pressure of the water, we use the term pounds—there is 24 pounds pressure to the square inch, say. The volt takes the place of pounds where the current of electricity is concerned, and the term to indicate the pressure is "——— voltage." As the quantity of water that passes through a pipe depends upon the pressure, so the amperage of a current depends upon the voltage. In a battery the voltage should be about 1½ volts per cell. It is measured by a voltmeter.

Ohm is the unit of resistance. The wire, or other conductor, over which the current passes offers a certain amount of resistance, dependent upon material, diameter and length. Copper has less resistance than iron or steel. This is the reason why it is generally used in all cases where the length required does not involve too heavy an expense. The ohm measurement is taken by an ohm meter.

Watt is the unit of electrical activity, the equivalent of horse power as applied to a steam or gas engine. An electro-motive force of one ohm and a current of one ampere constitute a watt. As in the other cases, there is a specially constructed device for recording the number of watts.

CHAPTER XIV.

TIRES AND THEIR PROPER CARE.

The subject of tires and their care is an important one because they are the most expensive part of the motorcycle that actually wears out, and must be replaced. Motorcycle tires as a rule are of 28-inch wheel diameter and 2½ inches casing diameter, designated in the trade as 28x2½. The average cost of a set (two) of these tires is from $28 to $30, according to the "tread" finish—the

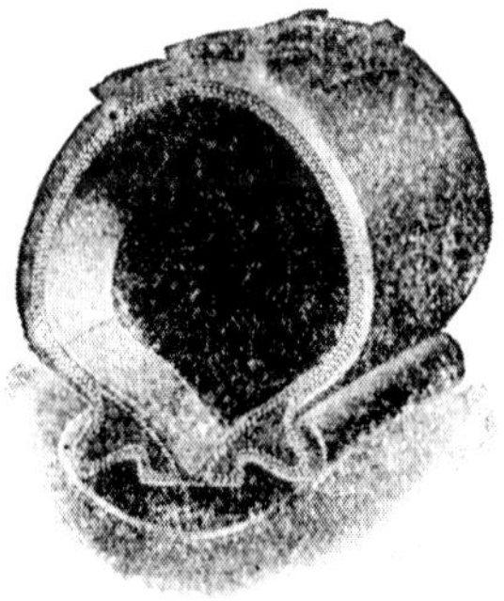

Interior of G. & J. Tire.

tread being the surface which comes into contact with the ground. For a long time all tires were made with a smooth surface, but, to lessen the tendency to slip, or "skid," most of them are now molded with raised projections of some kind on the tread. These have a tendency to hold to the road more securely than a smooth surfaced tire and in this way obviate or greatly lessen the chance of skidding.

Preference Given Pneumatics.

Nowadays all motorcycle tires are of the pneumatic —or air-filled—variety. Solid rubber tires are undoubtedly the most serviceable. They not only last longer than the pneumatics, but are immune to punctures and blow-outs. But there is one fatal objection to them—they have little resiliency, or spring, and the rider has to take a lot of hard jarring and pounding which is avoided by the use of pneumatics. The resiliency of the latter gives them a desirable rebounding, or springing quality similar to that shown by a rubber ball when thrown

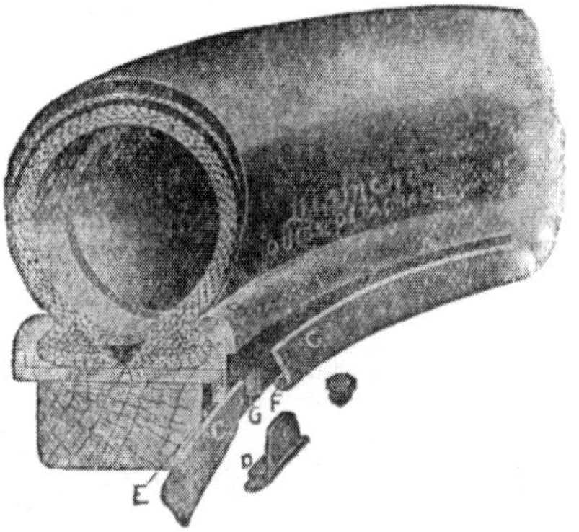

Sectional View of Clincher Tire.

against the ground. The rider, for instance, who runs over a stick or stone on a pair of solid tires gets the full force of the jar, less what is absorbed by the seat spring. When a pneumatic-tired wheel meets the same obstruction the tire gives enough to deaden the shock and make it hardly perceptible. This is one main reason why pneumatics are preferred—they ride easier.

For the same reason various substitutes for air-filled tires have not become widely popular. Some of these are all right in theory, but they don't work out satisfactorily in practice. Take the matter of tire "fillers". These "fillers" had quite a run a few years ago, but are not so widely used now as they were at the start, which in itself is a pretty fair indication that they don't exactly fill the bill. "Fillers" are made of a substance like that used in printing press rollers, a mixture of molasses, glycerine and glue. This is melted and put in the inner tube under pump pressure. When the substance cools and hardens there is no danger from punctures as there is no air to leak out. The very absence of this air, however, does away with the requisite resiliency and when the

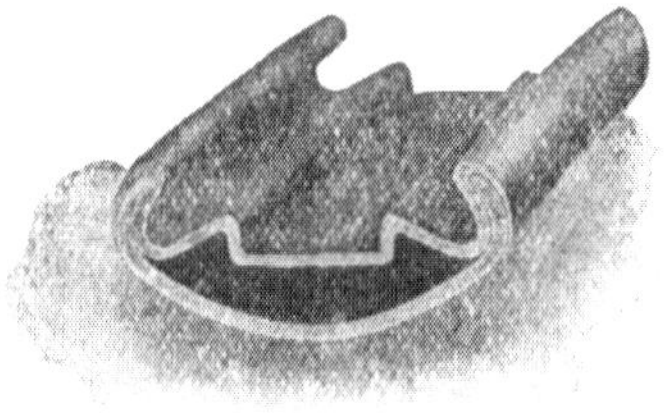

Section of G. & J. Clincher Rim.

rider attempts to move his machine at a fair rate he gets a severe jolting. The "fillers" are also affected by climatic conditions to some extent. In hot weather the filling softens and if the wheel stands in one position for any considerable time that part of the tire which touches the ground becomes "flat," and more uneven riding ensues.

How Long Should Tires Last?

"How long will a set of tires last?" is a question frequently asked. No positive answer can be truthfully given. As a general proposition, manufacturers will say that 4,500 miles is a fair average distance for a set of motorcycle tires to travel, and some of them will guarantee their product to give this amount of service. No maker will attempt to name any definite length of time for which tires may be used, as this is governed solely by conditions. One rider may wear out a pair of the highest grade in a month and travel only 2,500 miles, while another will make a set of inferior make last him a year, covering in that time 4,500 miles, or even more.

The nature of the roads travelled is a great factor in the wear of tires. If the roads are stony, or covered with particles of sharp flint or sand, the wear will naturally be greater than it would be on smooth, soft earth. Another determining factor is speed. There is a lot more destructive wear in tires when the machine is moving rapidly than there is when an ordinary gait is maintained. This fact holds good regardless of road conditions. There is bound to be more or less friction between the tire surface and the ground, and as the tire is made of a destructible material, this material must in time show the effects of the friction.

Then there is still another factor in tire wear; it is weight. A rider of 250 pounds weight holds the machine down harder to the ground than one of 150 pounds, and accordingly the frictional wear on the tires is increased by just this difference in the weight pressure.

Keep Tires Well Inflated.

Well-inflated tires will give longer and more satisfactory service than those which are not properly inflated. Novices are almost always afraid of over-inflation and the result is they seldom get in air enough. If the tube is not properly inflated the tire will cut out rapidly at the rim. Every motorcyclist should have a hand pump with a gauge so he may know definitely the exact amount of tire pressure. On tires 28x2½ the proper pressure for the front tire is 45 pounds, and on the rear tire about 50 to possibly 55 pounds. Few hand pumps will give a pressure of 100 pounds to the square inch, and as good tires will readily withstand this and more, there is no danger of over-inflation. The reason for putting more pressure in the rear tire than in the front one is that it has a heavier load to carry. Tires thus inflated should sustain a weight of 225 pounds to the wheel.

New tires should be inflated frequently during the first ten days or two weeks of use. When a tire is new the casing has more or less give or stretch to it. For the first ten days or so there is a slow, but persistent expansion of the casing under the pressure of the inflation and as this expansion continues the pressure is proportionately lessened.

Lack of Inflation Causes Tire to Flatten.

Effect of Turning Corners at High Speed.

This expansion and lessening of pressure in some instances gives the tire the appearance of partial deflation, and the beginner, not knowing the real condition, is inclined to believe there is a leak. When the expansion period is over, which will be in about two weeks, or a little less, inflation every ten days should be ample. If it is required oftener look for a leaky rim valve, or a slow puncture.

Hints on Care of Tires.

Inner tubes will wear better if they are liberally treated with French chalk. This may be done by using an ordinary bug-powder blower, and squirting the fine chalk powder between the tube and the casing. Where a blower is not to be had the chalk powder may be rubbed on the tube by hand before it is inserted in the casing.

A layer of felt, cut so as to be a trifle narrower than the tread of the tire, and inserted between the inner tube and the casing, will tend to lengthen the life of the former and lessen the liability of puncture.

Whatever is done in the way of repairs to tires by the roadside must of necessity be of a temporary nature. The only permanent relief is in vulcanizing the injured part.

Some cements contain acids. These act promptly and satisfactorily so far as obtaining quick sticking or adhesive results is concerned, but are death to tires. The acid in time rots the tire, not only making it more susceptible to blow outs, but rendering impossible any permanent repairs by the vulcanizing process.

Some Enemies of Rubber.

Oil, gasolene, water, extreme light and heat are all injurious to rubber. Oil and gasolene are solvents for rubber and will destroy it quicker than any other substance except acid. Light and heat exert a chemical action which causes speedy deterioration. This deterioration is slow, and unnoticeable, but it may be materially lessened by keeping your machine as much away from extreme light and heat as possible. In storing a machine for the winter this may be accomplished without trouble. When you stop by the roadside during a trip on an extremely hot day, put the machine under a shed, under a tree, anything that will keep off the sun's rays. Your tires will last much longer.

Water rots the canvas in the casing, rusts the rim, and weakens the rubber. Every man who uses rubber tires must have noticed they cut and puncture more readily when wet than they do when dry. Try to cut a piece of rubber with a dry knife. Then wet the blade of the knife, use it on the same piece of rubber and note the difference.

While the heating of tires is primarily caused by the friction of the outer casing on the ground, there is some little play, and resultant friction between the inner tube and the casing. To lessen this, dust the tube well with French chalk. This is the only satisfactory lubricant for rubber.

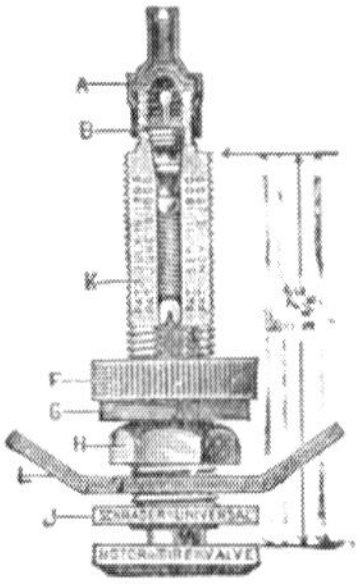

Air Valve for Clincher Tire.

A little care exercised in picking one's way along a bad road will tend to lengthen the life of the tires. It will be time well spent. It is needess, inexcusable extravagance

Start slow—There should be no exception to this. If you wish to ride fast wait until your machine is moving well, and the road is clear. The shock of a quick start strains the tire severely.

Take curves and corners at a moderate pace. If you to ride heedlessly over rough roads or through a litter of broken glass and similar obstructions. And yet many oft-times beyond repair.

unthinking men do it, cutting and damaging their tires

swing around a curve or corner at high speed the side strain thus created will make the machine "skid," and also tends to tear the tires from the rims. In either event a serious accident may result.

Keep your machine out of ruts in the road. When you ride in a rut, particularly if the ground is hard, such as baked clay for instance, the sides of the casing next to the rim will be cut through, and the rim itself may be badly bent.

Working of the Clincher Tire.

Tires of what is called the "Clincher" pattern are used exclusively on motorcycles. This is solely because they may be gotten at for repair purposes without taking off the wheel. In the case of the motorcycle the taking off of a wheel is a serious proposition as it involves an almost entire dis-assembling, or taking apart, of the rear mechanism of the machine.

Clincher tires are held to the wheel rim by air pressure exerted through a clinching process, hence the name, When the air pressure is released by deflating the tire, this "clinch" is loosened and the tire, or any part of it, may be raised from the rim. A glance at the accompanying illustrations will make the working of the clincher tire plain. See page 162.

The edge of the inner tube, which is always of thin rubber, is shown slightly back of the opening. Next comes the casing, which consists of alternate layers of heavy fabric and rubber composition. These may be distinguished by the difference in texture shown in the cuts, the fabric being speckled, and the rubber composition solid black, the fabric acting as a reinforcement. Pure rubber, or anything approaching pure rubber, would be out of the question in casing construction for two reasons. In the first place the expense would be too great, and in the second place rubber would not afford the necessary resistance to wear. Hence a composition of rubber and various other materials which tend to harden it, and impart wearing qualities, is employed.

At the lower part of the casing it will be noticed that there is a surplusage of material, more than is needed to form the tubing of the casing itself. The wheel rim is made with rounded edges bending inward. When the tire is deflated it is easy to push this surplusage of the casing under the bent edges of the rims. Then, when the air is applied and the tubing inflated, the entire tire,

Typical Anti-Skidding Tire.

casing and all, swells out so that the lower part, what we have arbitrarily called the surplusage, is gradually forced out to fill the vacancy in the rim surface. When the limit of proper inflation is reached the tire is thus held fast, or "clinched" so securely that nothing short of an unusually hard wrench, like that imparted by "skidding" at high speed, will tear it loose. At the same time it is readily removable when the air pressure is relieved.

Pure india rubber, the caoutchouc of South America, will float in water just as cork does. Little, if any absolutely pure rubber is used commercially, however, although it is customary to refer to inner tubes as "pure rubber." Everything in the rubber line which enters into the make-up of a tire is more or less vulcanized, hardened with sulphur to give it wear resistance. Inner tubes come in two colors, red and gray. The red has been treated with some metallic oxide, generally antimony, which gives it strength, but at the same time diminishes its elastic properties. The gray tubes contain merely the three percent of sulphur used in the original vulcanization process. The elasticity of the latter is therefore preserved, and a gray tube will usually stretch seven times its length, while the limit of the elasticity in a red tube is not more than five times its length, and not that if over fifty percent of oxide has been used. Here we are up against a peculiar proposition; which is the best, the unadulterated soft, elastic gray tube, or the adulterated, but stronger red one?

G. & J. Outside Tire Sleeve.

G. & J. Inside Tire Sleeve.

CHAPTER XV.

COST OF OPERATION.

What does it cost to run a motorcycle? Here again we have a question in which the personal characteristics and habits of the rider are of vital importance. Some men are naturally careful and frugal; others are heedless and extravagant. These traits stand out just as strongly in the handling of a motorcycle as they do in the personal expenditures of the individual. One man will waste gasolene by continually using too rich a mixture, and by allowing his engine to run at top speed when there is no necessity for it. This increases his gasolene bill. He will waste oil by over-lubrication. He will ride pell-mell over rough roads, cutting up and otherwise injuring the tires, and frequently breaking some delicate part of the mechanism. All this costs money that might easily be saved.

Another rider, under precisely the same conditions, will avoid useless expense by using a gasolene mixture of proper proportions, and stopping the engine whenever occasion will admit of it; by regulating lubrication so there is no waste of oil, and by exercising a little care in the manner in which he moves over bad roads.

These instances represent the two extremes of wastefulness and economy, and also a wide margin in money outlay. It is practically impossible, therefore, to fix upon any basis of maintenance or operation cost that will hold good in all cases. The best that can be done is to state the average expense which an ordinarily careful rider will have to meet, leaving out of the calculation the item of repairs. Accidents will occur at times, even when the greatest of care is exercised, but this care reduces the likelihood of their occurence.

Cost of Gasolene and Oil.

Under ordinary conditions one gallon of good grade gasolene will be sufficient for fifty miles of travel. The price of this varies according to the locality in which it is bought. In some places it may be had as low as 15 cents a gallon, while in others the price is 20 cents, and occasionally a little more. A fair average is 18 cents. This would make the gasolene expense 18 cents for each fifty miles of travel.

Lubricating oil, like gasolene, varies in cost according to where it is purchased. It ranges in price from 90 cents to $1.25 a gallon. An average of $1.00 will not be far out of the way. One-half pint is more than sufficient for a run of fifty miles under normal conditions. At $1 a gallon this would cost 6¼ cents.

This gives 18 plus 6¼, or practically 25 cents as the fuel and lubricating expense per fifty miles of travel. For 100 miles the outlay will, of course, be double, making a cost of ½ cent per mile for these items, which is cheap travelling.

One Man's Experience.

In this connection the account kept by M. A. Cordner, a New Jersey rider, as published in Motorcycle Illustrated, is of interest. Evidently a professional bookkeeper, this man itemized every outlay on a trip of 3,950 miles, including the depreciation in the value of the machine incident to the wear and tear of the long trip. The account stands as follows:

Total cost, machine included, per mile	.0404
Total cost of repairs and accidents, per mile	.0015
Total cost of gasolene, per mile	.0038
Total cost of oil per mile	.00082
Miles per gallon of gasolene	52.666
Miles per gallon of oil	1316.6

Average cost of gasolene per gallon........ .2055

His itemized account gives the total cost of the 3,950 mile trip as $159.94, of which $100 is charged up for depreciation in the value of the machine, leaving $59.94 as the actual cash outlay, 1½ cents a mile, including repairs and cost of outfit worn or used by the rider. Deducting this latter item—$31.15—because it has no proper place in the operating expenses, we have $28.49 as the real expense of the trip, less tire cost, or 75 cents per 100 miles. The way Mr. Cordner figures it is this:

REPAIRS CAUSED BY ACTUAL WEAR.		
Rivets, bolts, nuts, and retainer ....$	.50	
Lining for band brake	.50	
Two short drive chains	3.20	
Two compensating friction rings	.90	
Small engine sprocket	1.00	$ 6.10
REPAIRS CAUSED BY ACCIDENT.		
Repairs to cyclometer	.25	
Bolt	.10	
Hand made nut for engine sprocket	1.00	
Vulcanizing rear casing	1.75	
Control nut and yoke	1.05	4.15
OUTFIT.		
Clothing, outfit, speedometer, luggage carrier, spares, etc.		$31.15
		$41.40
Gasolene, 75 gals	15.29	
Oil, 3 gals	3 25	
Depreciation (market value)		100.00
Total cost, machine included		$159.94

One and One-Tenth Cent a Mile.

It is evident Mr. Cordner has figured the tire expense in the depreciation of the machine value, as in a long trip this item would be heavier than any other, and yet it does not appear in the account. Tires cost about $30 a set and are calculated to withstand the wear of from 4,-000 to 5,000 miles of travel. In this instance, therefore, they might as well be put down as a three-quarters loss, a further charge of $22.50. As there will be a salvage of $7.00 on the inner tubes the net tire expense is $15.50; this brings the total actual operating outlay up to $43.90. Calling it $44 even, we have one and one-tenth cent per mile.

There is no excuse for charging the entire cost of the clothing, speedometer, luggage carrier, etc., as items of expense on this one trip. The speedometer and luggage carrier are retained, and are of full value for subsequent tours. The matter of special clothing is one of luxury and comfort, governed solely by the rider's desires. There was no actual necessity for it—ordinary garb would have answered just as well, and this would have been worn if the trip had not been made.

There are few people who make 4,000-mile tours. The average run for business or pleasure is rarely over 100 miles. Hence the item of tire expense which figures up pretty strong on a long tour, is reduced very materially in the ordinary usage of the machine.

It is a safe proposition that, under normal conditions, the expense of operating a motorcycle will not exceed one cent a mile, barring the cost of repairs. What this latter item will be no man can predict. There are other essential items in the cost of operation which may be figured out with certainty. These cover gasolene, oil and tires.

CHAPTER XVI.

EFFECT OF WEATHER ON MOTORCYCLES.

Strange as it may seem climatic conditions have a marked effect on the operation of a motorcycle. Cold weather, for instance, will affect the gasolene and oil. Gasolene vaporizes slowly on a cold, damp day, and consequently requires less air to ensure a proper working mixture. It is not unusual for the vaporization to be so backward that the motor is provokingly slow in getting into action. Cases are also known where a motor worked satisfactorily one day. During the night a heavy cold rain came up, continuing until the next day. The motor got thoroughly chilled, and "balked" when the rider tried to start it. When it did start it was with a jerky, intermittent action. The rider was mystified because the adjustments had not been changed in any way. It was simply a case of slow, insufficient vaporization. As the motor warmed up the action was a little better, but it was far from being satisfactory until the amount of air admitted to the mixing chamber of the carbureter was reduced.

On the other hand, on a dry, hot day the vaporization is more rapid, and, therefore, more air is necessary in order to keep the mixture down to the proper proportion.

Cold Weather Affects Oil.

Oil will congeal on a cold day and thin on a hot one. Occasionally the congealment will be such that it is difficult for the oil to work its way through the feed pipe, and sometimes it becomes so thick as to clog the pump. In a case like this it will be found well to heat some of the oil in a can by placing it on a stove. Then just before you are ready to start pour this warm oil into the oil tank. It will thin the congealed oil in a few minutes. In lieu of this the oil tank may be wrapped in towels or flannels which have been soaked in hot water. When this latter course is followed it is sometimes necessary to repeat the application of the towels several times before the oil is thinned properly.

Dampness Causes Mis-firing.

Wet weather will cause the motor to mis-fire. It is a prolific cause of ignition troubles. If water gets to a wire through a worn, or otherwise imperfect insulation, it causes an electric leak which weakens the current so that the action of the spark is irregular, and mis-firing ensues. The same wire, when the insulation dries out, performs its duty satisfactorily; and there is no more trouble until it gets another wetting. This is also a mystery to the novice, as the leak is difficult to locate unless the rider chances to pass a moist hand along the defective wire, in which case he will get a shock that will put him wise to the location of the trouble.

Water is also injurious to magnetos, interfering with the flow of the current. Realizing this experienced riders use an oilskin or rubber cover with which they envelope the magneto to protect it from a heavy rain.

Water Weakens the Tires.

And above all, don't forget that water is a destroyer of tires. When caught out in a heavy rain go slow and pick your way over the road carefully, choosing the smooth places. Tires cut and tear, and develop punctures much more readily when wet than they do when dry. Obstacles that can be passed over without damage when the tires are dry inflict serious rents and tears when they are wet.

CHAPTER XVII.

SELECTION AND USE OF MACHINE.

In picking out a new motorcycle be sure that it fits you. This is more essential than that your clothes should fit. The latter is merely a matter of satisfaction and personal pride, while the former is one of actual comfort and pleasure. A person abnormally long of legs or arms, for instance, would not be comfortable on a machine the frame of which properly accommodates a man much shorter in legs and arms. Get into the saddle and

One of the New Seats.

make sure that your legs rest on the pedals or footboards without stretching or cramping; that your hands can grasp the handle grips without undue effort of extension or contraction; that you are not called upon to bend out or draw in your arms or hold them in an unnatural position. The saddle should be of a height that will enable you while sitting on it to place both feet firmly on the ground so as you may protect yourself from a "skid," or stop momentarily without dismounting. All this, it is true, may be secured to a degree at least by raising or lowering the saddle, but it is just as well to have the latter in the right place to start with, and besides there are limits to the saddle range.

See that the saddle is placed where you will have an easy position when riding. And above all make sure that it is of the right form and make. You will find this an important item on a long journey.

The Matter of Cylinders.

It matters little which particular make of machine you buy. All the standard makes are good. Every manufacturer, for selfish business reasons if nothing else, is trying to do his level best in turning out a machine that will give satisfaction. The principal choice to be made is in the number of cylinders, and in this you should be governed by the use for which the machine is intended. The "torque," or turning action of the crank-shaft is a little more regular and continuous in a four-cylinder machine and hence the vibration is less, but the one and two-cylinder machines now on the market are also practically vibrationless, great advance having been made in the placing, as well as in the operation of the engines. Multiplicity of cylinders certainly gives increased power and this is a valuable feature if there is much hill climbing to be done, or the roads are unusually heavy. If you are a rank novice, and have never handled a motorcycle, it is altogether likely that the dealer will advise you to begin with a one cylinder machine as there is not so much mechanism to look after. Then, as you become proficient you may take up the two-cylinder outfit, and finally the four. This is most sensible advice.

Other Vital Points.

There are other features, not of a mechanical nature, which are of more vital importance than the question of the number of cylinders. These are:

Where is the factory located? Is it within a reasonable distance so there will be no delay in obtaining new parts if they are required?

Is there a dealer near you to whom you may apply for help in case of trouble?

Does the manufacturer guarantee his machine to give a certain amount of wear, and is the guarantee good?

What about repairs which may be necessary, through no fault of yours, before the guarantee period expires?

Having received the machine, and being ready to take your initial ride on it, first fill the gasolene tank, and next

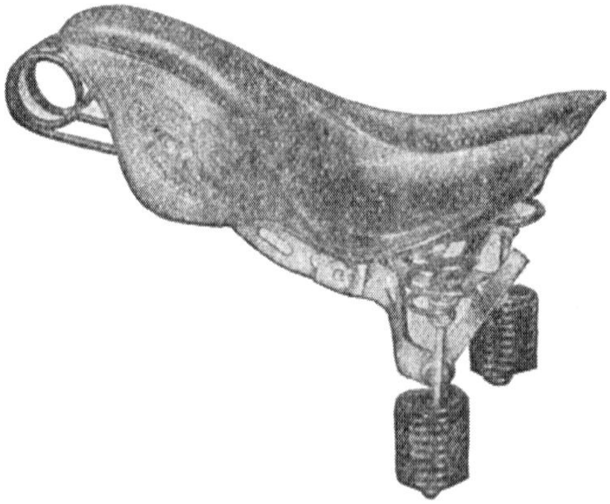

Typical Motorcycle Saddle.

the oil tank. See that the crank case (if the "splash" system is used) contains sufficient oil. Bear in mind the fact that the new machine will require more oil than one which has been used some time. Put the spark plug in place, connect it with the sparking wire, and you are ready to start. This will require a little pedalling, or cranking, (according to the style of machine you have bought), in order to secure the compression necessary to set the engine in operation. Leave the carbureter and magneto or battery, and similar parts alone so long as they work well. They have been put in proper condition at the factory. With each machine comes a set of printed instructions which will be your guide in connecting such parts as may need your attention.

Start slow and run slow. Don't expect too much of your machine in the beginning. It will run better after being used awhile, partly because whatever rough edges may possibly have been overlooked at the factory will wear off, but more largely because you will have become better acquainted with the mechanism, have more confidence in yourself, and know how to proceed more intelligently.

Don't Tinker Too Much.

Repress the tendency to tinker with parts of the machinery which you don't understand, and which really need no attention. Many a good motorcycle is put out of commission by this tinkering process, which in the great majority of instances is inexcusable. Familiarize yourself with the various parts of your machine, and learn the names by which they are designated. This you may do with the help of the printed instructions furnished by the manufacturer. Many of these contain a "keyed" illustration in which the various parts are numbered and then described in the text as is done in the case of the Excelsior machine in this book. In this way you will readily become familiar, not only with the parts themselves, but get a clearer idea of the function each performs in the operation of the whole.

Taking the First Lesson.

If you have ever ridden a bicycle the mounting of a motorcycle will be easy as the balance (equilibrium) of the wheel is much more readily maintained. In this respect, the "safety" (low-wheeled) is a great improvement over the "ordinary" (high-wheeled) bicycle, but even in the former the weight is almost entirely above the line of the wheel hubs and consequently the rider must frequently perform an act of balancing by swaying his body to one side or the other in order to maintain an

Rear Wheel Raised With Stand.

equilibrium and keep the machine from falling over. In the motorcycle the weight is much more evenly distributed and it requires very little exertion on the part of the rider to keep the machine on an "even keel."

When the owner of a new machine has never ridden on a bicycle or motorcycle it will be well to take a lesson or two in the privacy of the barn or garage, or in a yard, before venturing on the road. This may be done to the best advantage in the following manner:

Running around the rear wheel, hinged to the axle on both sides, and held up with a hook to the rear end of the mud guard, you will find a metal arrangement. When this is released from the hook and allowed to touch the ground under the wheel it acts as a stand, raising the wheel from contact with the ground, and holding the entire machine in an upright position. Unhook this stand and push it into place so the wheel revolves without touching anything. Ascertain from the manufacturer's instructions which way the handle grip must be turned to open the exhaust valve and release whatever compression may be in the cylinder. Then open the valve by turning the grip in the proper direction. This will enable you to move the pedals easily. So long as the exhaust valve is closed pedalling will be hard because of the compression it generates.

How to Mount and Pedal.

Now mount the machine. It will be difficult to overturn it on account of the support given by the stand under the rear wheel, and any ordinarily active man may get into the saddle without difficulty. Get on and off a number of times so as to familiarize yourself with the operation and beget confidence. This secured take a little experience in pedalling which is really a most simple operation and yet occasionally novices find it confusing. The one requisite is to move the feet alternately, one pedal going down as the other comes up, and to apply the force (the foot pressure) on the downward stroke. Some beginners, however, find themselves possessed of an almost unconquerable tendency to move both feet at the same time, but the great majority find the required movement easy as it merely entails the same motion used in walking, with a pressure added at the beginning of the downward stroke.

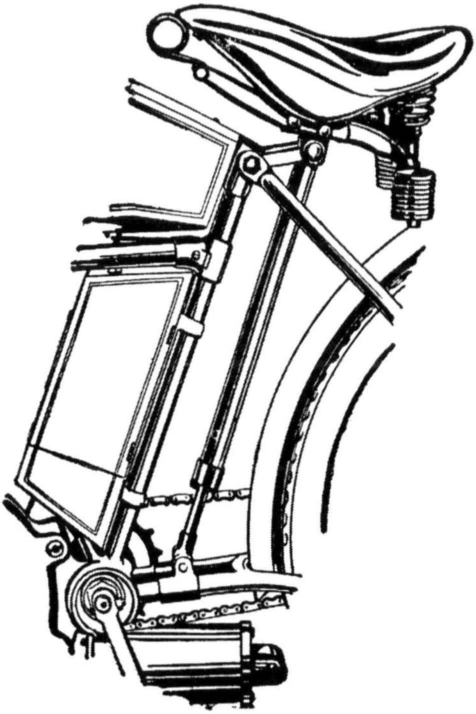

Greyhound Shock Absorber.

As all motorcycles, with one exception, are equipped with pedals which must be used to generate compression to start the motor with, as well as a means of getting home in case of emergency, it is important that the rider should know how to pedal properly.

No matter how timid, or awkward, or uncertain a

novice may be, an hour of this kind of exercise will beget confidence, and he may then practice steering until he knows without stopping to think, that turning the handle bars to the left turns the front or steering wheel in the same direction, and vice versa.

Learning How to Steer.

Having mastered the mounting and pedalling problems the beginner should put the stand back to its road position by hooking it to the rear mud guard. This will leave both wheels on the ground. Get some suitable support, a stout box of the proper size will do, and place it under the forward part of the frame so as to lift the front wheel from the ground and allow it to move freely. Get back into the saddle, allowing the feet to drag by the side of the machine, or rest them on the pedals if you can overcome the impulse to move them. Remember that the rear wheel is now resting on the ground and that if you move the pedals the emergency drive will be set in motion propelling the rear wheel forward and moving the machine while you wish it to remain stationary. This, of course, could be prevented by leaving the rear wheel raised on the stand, but unless the front support is very solid and well placed there would be a "wobbling" motion which would distract and annoy the novice, taking his mind from the job in hand. Take hold of the handle bars lightly but firmly. Don't grasp them with a death-defying wrestling hold, but learn to graduate your grasp so that a sudden jar or shock will not break it, and at the same time it will not be tense enough to tire you wrists and arms. Too tight a grasp also incites a tendency to unwittingly turn the handle grips while merely trying to swing or move the handle bars themselves. When the handle bars are moved to the right or left the front or steering wheel moves accordingly, but when the handle grips are turned the control of the engine, the regulation of the valves, the proportions of the gasolene mixture, the position of the throttle, etc., are also affected.

At first there will be a natural tendency to overdo the steering act; to move the handle bars too far. When these are held exactly even the front wheel is in the same position, and the machine travels straight ahead. The slightest variation in the position of the handle bars either to the right or left produces a similar change in

Typical Double Grip Control.

the position of the front wheel and the course of the machine is changed accordingly, but in a greatly increased ratio. Thus a very slight movement of the handle bars will deflect the course of the machine a foot or more. Contrast closely this response of the steering wheel to the handle bar movement and you will quickly learn the trick of steering.

On the Road at Last.

The man who takes this preliminary practice will find the time to have been well expended when he comes to take his first ride on the road. Mounting and dismounting, pedalling and steering, awkwardly done by most novices, will all come easier for him. He will have a confidence which will secure rapidity of thought and action which is vitally essential when actual road emergencies arise.

Supposing that the gasolene and oil tanks are filled, the spark plug in place, and the wire connections made, the rider is ready to start. In the first place he should retard (slow-up) the spark. Then get into the saddle and pedal to secure the necessary compression. When the motor is in fair operation let a little more air into the carbureter by opening the air intake, through the handle-grip or lever control. From then on it is merely a matter of regulating the gas supply and the sparking. On level roads cut the gas down as low as possible and, if more speed is wanted, advance (quicken) the spark, and open the throttle wider.

On a hill where power rather than speed is needed, more gas must be used, and as the pace decreases retard the spark. If the hill is a severe one give the motor a full supply of gas; if the motor should labor hard, or show signs of failing, give the pedals a few sharp strokes. When a level is reached again cut down the gas and advance the spark gradually until you are moving at the speed desired. Aside from these hints, and the proper manipulation of the steering gear there is little to be learned except by actual riding experience.

When a stop is necessary shut off the gas by closing the throttle. Without fuel the motor will not run. If under high speed at the time, its momentum may carry it beyond the desired stopping place. Should you desire to avoid this apply the brake.

Should an accident or unexpected stoppage occur, consult the chapter on road troubles.

The Matter of Speed.

All standard motorcycles are capable of a sustained speed of from 35 to 40 miles an hour. This is fast enough for any sane man to ride for business or pleasure. It is too fast, except in case of emergency. Twenty miles an hour is a much more reasonable, and far safer pace, and the man who travels at this gait, or even slower, will get more comfort and satisfaction than the one who travels faster, because he will have an opportunity to see things as he goes along. His mind will be free from the care

The Veeder Cyclometer.

which fast riding compels, and he will have opportunity to take note of the scenery and otherwise enjoy his trip.

But, while the ordinary standard-make machine is only guaranteed for a limit of 35 to 40 miles an hour it is a well known fact that almost any stock machine can be speeded up to 60 miles an hour on a good level road.

Don't worry about the speed qualities of a machine unless you are going into the racing game. Then you you will find extreme speed is almost solely a matter to be regulated by the ratio of the drive. You may take a machine which, rigged as it comes from the factory, has a limit of say 40 miles an hour and by increasing the size of the pulley on the crank shaft, speed it up to almost any gait desired.

What the Cyclometer Does.

With this handy little attachment shown on page 187, a motorcycle rider may tell exactly how fast he is travelling, how far he goes on the trip he is then making, and exactly how much ground he has covered during the season. The upper dial records each mile travelled on a trip, and is set at zero on starting out. The lower dial gives the grand total and is left undisturbed for the season, so that at any given time the rider may tell to a fraction of a mile how far he has run. This is a device made by the Veeder Mfg. Co.

Acetylene Lamps Favored.

There is only one lamp to use on a motorcycle, and that is the acetylene. Oil is unsatisfactory as the light is too weak, the wicks require frequent trimming to keep them clear of the charred portion, and the oil tank must be refilled, a nasty job at the best, to say nothing of the difficulty of procuring the right kind of oil in out-of-the-way places. Electric lights such as are used on automobiles are out of the question as they require too much extra power over that generated by the magneto. Men who have had lots of experience prefer the acetylene lamp. This consists of a chamber containing calcium carbide on which water drops from a small tank. This decomposes the carbide and releases a gas known as acetylene. The gas collects in a reservoir, and from there passes through a filter of horsehair, slag, or similar material which cleanses it of carbon. It next flows into the gas bag which equalizes the pressure so that when it reaches a certain stage the water is shut off, and the production of gas stopped. As the supply lowers the pressure is relieved, the water begins to drop again, and more gas is made. From the gas bag the gas flows into a metal tube at the top of which is a burner made of lava. This burner has two very small holes opposite to one another, and at an angle so that the two streams of gas meet and, when ignited, form a fish-tail flame. This flame, projected by a polished reflector in the rear, throws out a steady, intense white light which illumines the road for a long distance, and is not only a great help to the rider in steering clear of obstructions, but can be seen for a long distance by those coming from an opposite direction.

Lever in Coaster Brake.

In many places a tail light is also required. Here again the acetylene lamp is the best. In the case of tail lights the lens must be red so that anybody coming up from behind may know there is a vehicle ahead, and also know the direction in which it is moving.

The acetylene lamp is not liable to jar out when passing over a very rough road as an oil lamp will.

Useful Mathematical Rules.

Circumference of circle=diameter × 3.1416
Diameter of circle=circumference × 0.3183
Area of circle=square of diameter × 0.7854
Length of arc=number of degrees × diameter 0.008727.

To find the area of a triangle, as explained in the Automobile Trade Directory, multiply the base by one-half the perpendicular height.

To find the area of a trapezoid, add the two parallel sides together and multiply the sum by half the perpendicular distance between them.

To find the area of a regular octagon, multiply the square of the diameter of the inscribed circle by the decimal .828.

Reading Standard Motortricycle.

This is an entirely new form of motorcycle, most of those previously made having the two wheels in the rear. While designed to carry only two passengers, provision is made for a third by an extension seat which is bolted to the pan-like arrangement over the rear wheel. When in place this third seat projects to the rear so that the extra passenger is not in the way of the driver.

CHAPTER XVIII.

HANDY HINTS ON SHOP REPAIRS.

"How shall I do it?" is a question with which the man who wants to repair his own machine often has to grapple. Conditions are constantly arising in which there is no regular course of proceedure laid down in the text books, occasions when the motorcycle rider,

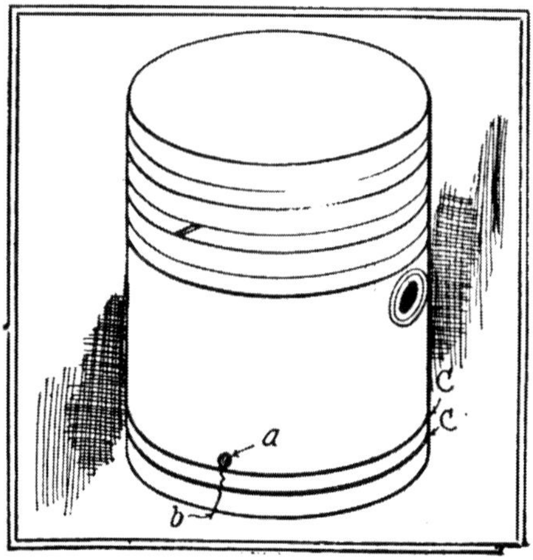

Crack in a Piston.

who takes almost as much pleasure in making his own repairs as he does in riding his machine, must act of his own inventive initiative. It is for the guidance of such that the following "shop hints" are given, the word "shop" being used advisedly as the repairs referred to are mainly of a nature which preclude their being made by the roadside.

Should a crack appear in the cylinder, drill a small

hole at the extreme top and bottom ends of the crack, going a little beyond the end of the crack in each instance. Insert in these cracks short "grubs," or headless screws, preferably of copper. These should fit snugly. If steel screws should be used the act of forcing them home might enlarge the crack. Copper, being softer, will accommodate itself to the space, and undue pressure on the defective part of the cylinder is thus avoided. Over the crack, and extending well beyond it in all directions, place a layer of asbestos, coated with red lead. Where it is impossible to get asbestos, heavy wrapping paper soaked in lubricating oil may be used. On top of this packing place a thin sheet of metal of the proper size, and tap or drill screws through the metal sheet and into the cylinder so as to hold them together tightly, but be careful that the ends of the screws do not project on the inside so as to interfere with the action of the piston. A repair properly made in this way will often make it possible to use a cracked cylinder for some time.

Cracked pistons, which are caused by carelessness in handling, may be repaired in a somewhat similar manner, except as to the metal patch. As the piston fits snugly in the cylinder, there is no room for such a patch, if it were required, which it is not. The only purpose in making the repair is to prevent further spreading of the crack. This will usually be found on the lower edge, and is frequently difficult to locate definitely as to its limits. When this is the case, cover the part under suspicion plentifully with oil. Then clean off the oil and warm the piston. If there is a crack some of the oil will have worked its way into it, and the heat will cause it to exude along the entire length of the fracture. At this extreme end ,and just a trifle beyond, drill a countersunk hole. Into this hole lightly tap a copper rivet or plug. The countersunk part of the hole will prevent any projection that would scrape the wall of the cylinder, and the insertion of the copper plug will, under proper usage, prevent the crack from spreading. In a piston repair it does not make any difference if the end of the rivet reaches out on the inside of the piston, as there is nothing there for it to interfere with.

Removing a Piston Ring.

When it becomes necessary to remove a piston ring for any purpose it may be done the quickest and handiest in the following manner:

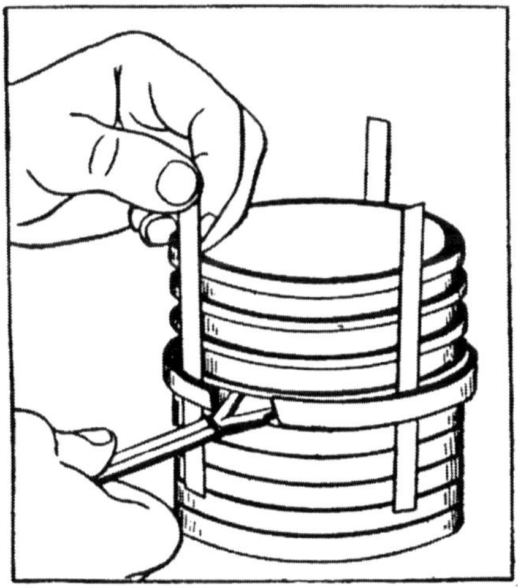

Taking Off a Piston Ring.
(Loosening the ring).

First remove the securing screw or ring which holds the piston in position, and then gently tap it out with a piece of soft copper or wood. Having thus released the piston have a companion hold it steady in an upright position on a table or work bench. You have previously provided yourself with three strips of tin about ⅜ inch wide, and 5 inches long. Next you need a pair of tongs or pliers made of ¼ inch diameter steel wire tied near one end with copper wire in such a way that the small or "tong" ends will open when the handles are pressed (closed) together. This is exactly opposite to the action of the regulation pliers. Insert the small end of the pliers in the ring slot and by closing the handles you expand the ring so that the tin strips may be slipped under it one by one, spreading them out so as to take up the pressure of the ring equally around the piston. (See illustration). Having inserted all three of the tin strips, drop the pliers, and remove the ring by the method seen in the second illustration, holding the ring with the fingers and pressing against the top of the piston with the thumbs. In attempting to remove a ring in the usual way, by forcing it apart to the limit of its expansion, and then dragging it off by brute force, it is liable to be broken or twisted out of proper shape. Besides this the method here outlined is quicker.

In replacing the ring exactly the same method (reversed) may be employed, the strips of tin being drawn out one by one when the ring is in position to snap back into place.

How to Solder Aluminum.

There is sometimes occasion to solder a broken aluminum or aluminoid part. It is a tough job for an amateur, and a lot of professionals hesitate to tackle it, but it may be done satisfactorily if the following directions are closely adhered to:

The best solder is composed of tin, 10 parts; cadmium, 10 parts; zinc, 10 parts; lead, 1 part.

First cleanse the broken parts thoroughly, by scraping if possible. Then allow them to stand for two or three hours in a strong solution of hypo-sulphate of soda. Or they may be immersed in a solution made of nitric acid with three times its bulk of hot water to which has been added 5 percent of commercial hydro-fluoric acid. Rinse the parts thoroughly on removing them from the solution or acid, and dry them in warm sawdust. In this way all grease or other substance fatal to a proper joint by the soldering process is taken off, and the surface brightended and sharpened as if it had been filed.

Naked aluminum when freshly scraped oxidizes rapidly on exposure to the air. This oxidization, although very thin, a mere film, prevents a perfect union of the broken

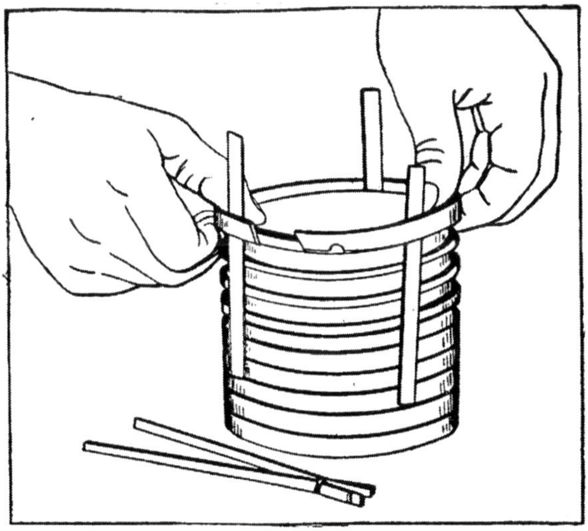

Taking Off a Piston Ring.
(Slipping off the ring.)

parts. To overcome this the operator must act quickly. Have the solder melted, and a nickel soldering tool well heated. Heat the parts to be joined, and scrape them well with the soldering tool while holding them under the surface of the molten solder. This not only tends to prevent oxidization, but the scraping with the hot nickel soldering tool removes such film as may have formed.

When the parts are well tinned in this way, and are of the proper heat, apply the flux, and bring the broken parts together, holding them in this position with reasonable pressure until a union is assured. The fluxing material may be nitrate of silver, resin, or zinc chloride.

Repairing a Valve Stem.

Broken exhaust valve stems are extremely rare, but an accident of this nature may occur, so it is well to know how to act. In the first place the stem must be removed. This may best be done in the following manner, as shown in the accompanying illustration on page 197.

Bend a piece of ordinary ¼ inch steel wire into the shape E, affixing a flat disc to the end which presses on the valve head, or the end of the wire itself may be bent in coil shape to serve this purpose. Next bend the other end into the form of a hook. Now take off the cover of the valve chamber and insert the coiled or disc end so it will rest on the valve head (see illustration). Next take a stout wooden rod or bit of metal. Pass this through the hook as shown at D, and insert one end under the valve spring A. Push down on the free end and the leverage thus afforded will enable you to easily overcome the spring tension C, so the cotter pin B which holds the stem in place, may be removed. This done, it is easy to lift out the stem, valve and all, through the opening over which the valve chamber cover fits, The principle is simple. The disc or coiled end of the wire which rests on the valve head holds the latter down snugly in its seating, while the pressure exerted by the lever D compresses the spring.

You are now in position to make the repair. When a break occurs it is generally at the point shown in figure 1, on page 199. Smooth off the rough ends of the break with a file. Next cut a piece of thin steel (G in figure 2) of the same diameter to a length which will give it a slight clearance between the valve in position and the tappet (H) at its lowest point. Then take a piece of steel tubing (F) which will fit over the tappet, steel binding piece and valve stem. Cut this to a length which will cover the entire repair. As the break and the filing has tended to shorten the valve stem the tension of the spring is accordingly lessened. To restore this to a normal state, cut a short screw thread on the valve stem (B) at the point shown by E. Next trim off the hooked end of the spring and put under it a washer

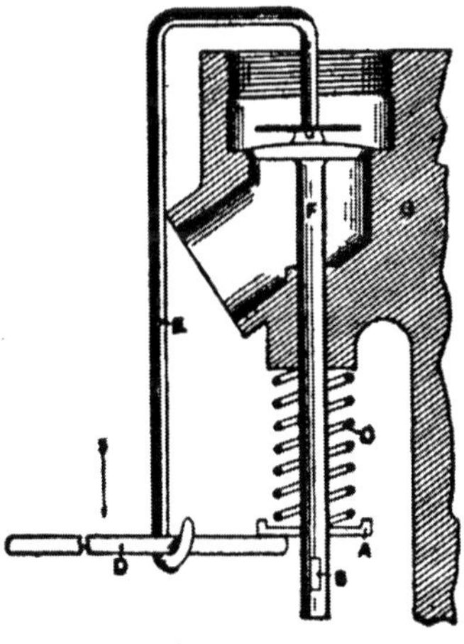

Removing a Broken Valve Stem.

(C), a nut and a thin lock nut (D) screwing this latter into the thread made at E. This restores the tension of the spring.

It will be cheaper and quicker to use a new valve stem, but in case of emergency where one is not to be had, good service may be gotten from a repair made in this way.

Repairs of Rubber Tires.

There is only one method of making permanent repair of a rubber surface, and this by vulcanization. This is effected by a combination of rubber, sulphur and heat, under presssure. Vulcanization may be accomplished satisfactorily with a crude outfit, the most important part of which is a hot flat iron, but the best way is to use one of the patent devices, a number of which are on the market.

The purpose in vulcanizing is to put a new, permanent surface on a torn or weak part of the tire. To secure this the rubber surface must be slightly melted so it will amalgamate with the added substances, thus forming a new and compact rubber surface.

Vulcanizing a Rubber Surface.

One of the first things to do is to remove the old patches, especially if they have been put on with cement for temporary purposes. This may be done with a flat iron which has been immersed in boiling water. Unless overheated, the iron may be placed directly on the patch and moved back and forth in the same manner a woman would use it in ironing. If there is a suspicion that it is too hot, merely hold it near the patch without actually touching it. This tends to loosen the cement when the patch may be peeled off without tearing the rubber of the tire. Next clean the surface with gasolene and when this has dried off, roughen the surface with sandpaper the same as in applying a temporary patch. Now apply a coating of rubber solution and when this is dry use crude rubber to fill the hole. Then put on the heat by applying the hot flat iron over a bit of cloth or canvas, leaving it for about thirty minutes at a temperature of about 250 degrees Fahr. (This may be done by using several irons and using them in relays so that the rubber will have no chance to cool). The heat thoroughly incorates the new and old parts. To effect proper vulcanization, three percent of sulphur is required, but there is generally enough of this in the rubber itself.

The rubber solution or cement mentioned is made of pure rubber (caoutchouc) dissolved in spirits, naptha or carbon bisulphide. It may be bought ready for use. Keep it tightly sealed as the spirit evaporates readily. Should the solution become stale it may be revived by the

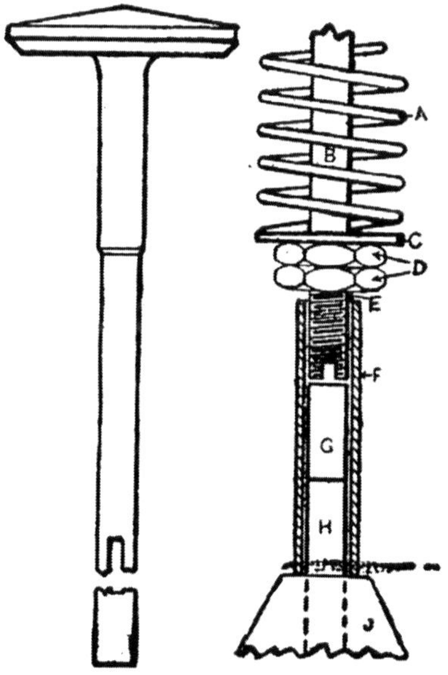

Fig. 1. Fig. 2.

Repair of Broken Valve Stem.

addition of a little spirits or naptha, but it is best to have it fresh. Keep it away from a flame, as it gives off a vapor which is dangerously explosive.

The same process is followed in the use of a regular vulcanizer, except that a hot flat iron is not necessary, the vulcanizer taking its place. In this the heat is generated by electric current.

Regrinding of a Valve.

Valves and valve seatings will wear unevenly at times and when this uneveness occurs the valves are sure to to leak, causing a loss of compression. When the valves are found to be in this condition the only remedy is regrinding. The simplest and easiest way of doing this regrinding is as follows:

Mix a little fine emery powder with lubricating oil, lift the valve away from the seating and apply the mixture to the face with one of your fingers. Put the valve back into place. If you are without a special grinding-in tool take a screw driver, put it in an ordinary bit brace and insert the end in the slot on top of the valve head. Then, with moderate pressure, partially rotate the valve backward and forward. Lift the valve occasionally and change its position in the seating so as to secure an even regrinding. Wipe away the oil and emery with a clean rag or waste. If the face of the valve and the seating are both bright and the valve "beds" down properly into position, the regrinding is sufficient. If any dull spots are left put on some more oil and emery and do a little more turning of the valve head.

Don't bear down too hard on the screw driver, or whatever tool you may use. Don't let any grit or foreign substance get mixed with the oil and emery. If you do uneven surfaces which can only be removed by turning in a lathe will be formed. When the job is done clean the valve and seating properly, using kerosene or gasolene.

When the valve is put back in place make sure that the lower end of the valve stem has a clearance of about 1-32 inch when the tappet rod is at its lowest position. It sometimes happens that the regrinding process lowers the valve and stem so there is no clearance. When this occurs file the top of the tappet rod a little. Don't touch the valve stem itself. If the valve is very badly worn or pitted, or out of true, it will not pay to regrind. The proper remedy is in a new valve.

Removing Broken Spark Plug.

One of the awkward jobs which has occasionally to be undertaken in a repair shop, says the Automobile Dealer and Repairer, is the removal of a broken spark plug. The break may be caused either by the brass part of the plug being firmly fixed, or owing to the head having been wrenched off in trying to remove it. In such a case, the tang (sharp) end of a file or similar tool will be found useful. A small hole should be drilled in the plug, and

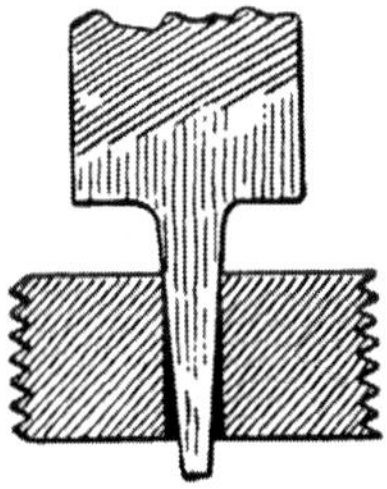

the tang end of a file driven tightly into the hole, the hardened end of the file being protected with a block of wood. The file should then be gripped with the jaws of a shifting spanner, and while turning the file so as to unscrew the plug, it should be kept pressed down very firmly into the hole. This is very often a successful way of removing a broken plug, but the file must be driven in firmly before attempting to turn it, otherwise the thread of the spark plug hole in the cylinder will be torn out.

Keeping the Hands Clean.

Overhauling a motorcycle generally begrimes the hands. Black dirt wedges itself under the finger nails and is difficult to get out, even by long soaking in hot water, while the oil stains on the hands themselves are hard to get rid of.

Before starting a job of overhauling or repairing, fill in under the tips of the finger nails, and around the sides where they join the flesh, with soap. This will prevent the dirt from getting into these places and, when the job is done a little soaking in warm water will remove the soap leaving the nails clean.

Lengthening a Piston Ring.

Sometimes a piston ring needs a little expanding. If you have a light hand and a little patience this expansion

may be obtained by a hammering process, despite the fact that the ring is made of cast iron. Set the large flat face of a hammer upward in a vise (see illustration) and screw it fast. On this place the piston ring, and with the peen (knob) end of a small hammer tap gently on the inside of the ring working from the middle both ways toward the slot. Go slowly and tap lightly or you will break the ring. Should the tapping expand the ring in width as well as in length, as it frequently does, it may readily be filed down with a flat file so as to fit the slot in the piston head.

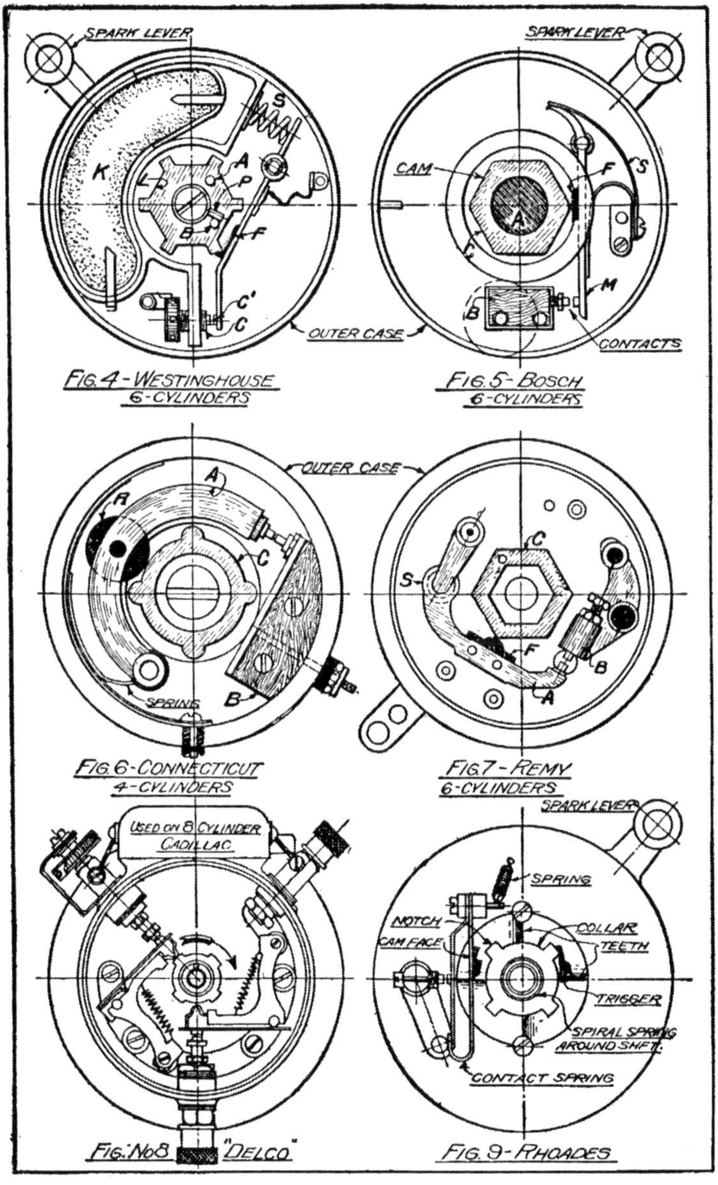

Figs. 4-5-6-7-8-9.—Showing Well Known Primary Circuit Breakers as Used on Modern Automobiles—See Detailed Description.

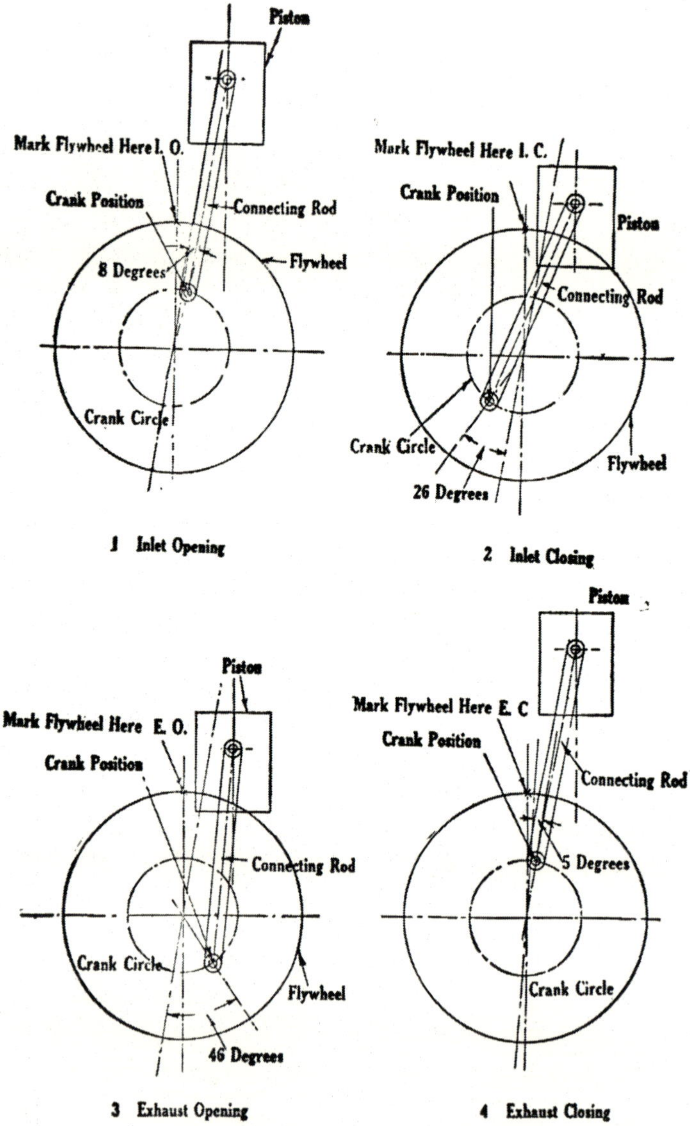

Diagram for Setting the Valves of an Offset Motor.
(Courtesy "Automobile Journal.")

CHAPTER XIX.

SOME THINGS WORTH KNOWING.

There is a limit to the carrying capacity of a motorcycle, hence it is impossible for a rider to have with him at all times everything for which he may on various occasions have use. There are some things, however, which he cannot very well get along without. Among these are:

Light monkey wrench and spanner.
Combination pliers.
Screwdriver and flat file.
Three spark plugs.
Tire repair outfit.
Spare valves.
Spare links (if chain is used).
Belt punch and fasteners.
Insulated tape.
Supply of insulated wire.
Oil can (filled).
Lock and chain.

This list may sound formidable, but it really occupies comparatively little space, and may be conveniently carried in a case strapped to some part of the frame.

Protection Against Thieves.

Should you have occasion to leave your machine unprotected by the roadside and have no lock and chain with you, take out the switch plug. This will prevent some chance stroller from riding off on it—unless he happens to have a spare plug in his pocket. The best way is to lock the rear wheel to the frame and also take out the switch plug. This gives a double protection.

Engine Testing Speeds.

When running your engine "on the stand" for tuning up, or testing purposes, don't let it race. Overspeeding is not only destructive to the motor, but makes it difficult to distinguish the very faults you are looking for. If there is erratic firing, for instance, it is hard to distinguish it if the engine is running at its limit as the explosions occur so closely together that an occasional misfire or pre-ignition might easily be overlooked. The best way is to retard the spark fully, then after a careful test at low speed, about one minute high speed should be all that is necessary. Care exercised in this direction will prevent many a road breakdown.

Keep Your Machine Clean.

Your machine will last much longer if you take decent care of it. When a railway engine comes in from a run "hostlers" take it in hand and give it a careful cleaning, while mechanics look over the mechanism to see that it is in working order before the engine goes out again. This is done because it pays in lengthening the life of the engine, and is also a safeguard against accidents. A motorcycle costs money—from $200 to $350, according to the number of cylinders, etc. Why not take decent care of it? The machine should have proper housing, and should be cleaned as soon as convenient after every trip, before moisture and mud have a chance to beget rust and other damages. Clean the cylinders occasionally with a charge of kerosene as explained in the chapter on pre-ignition. Keep the frame, wheels, and all exterior parts well wiped, and go over them once in awhile, say once a week, in a hunt for loose nuts and weak connections, and tighten them. If you use a chain drive, the gummed oil or grease may be removed by soaking the chain in kerosene. If you lay up your car for the winter it will pay you to take off the wheels and store them away in a moderately cool, dry place, safe from oil or water or acid. Preservation of the rubber tires will be further insured if the wheels are enclosed in a heavy green cloth covering, or placed flat in a well covered box, in the center of which is a saucer of turpentine. The fumes of the turpentine prevent oxidization of the rubber.

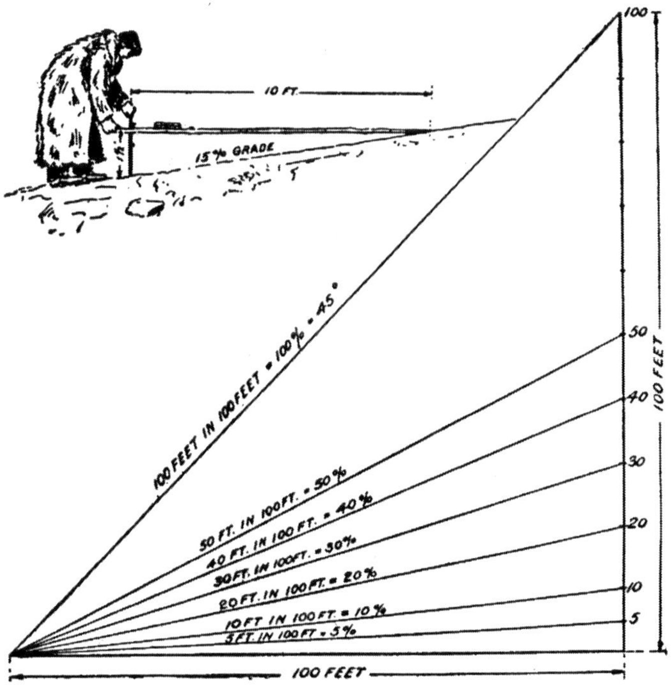

Handy Method of Measuring Grades.

Measuring Grades Without Instruments.

There are times when it is a matter of great satisfaction to know the exact grade percentage of a hill. This frequently happens on club or party runs when discussion arises as to the "steepness" of some rise which has been climbed. As it is not the practice of motorcyclists to carry gradometers, it is usually impossible to settle the question satisfactorily. And yet there is an easy method of doing so. It is as follows:

Take a straight rod or flat stick and measure on it ten feet, making a plain mark at the exact point. Place one end of this stick on the rising ground, and hold up the other, at the 10 foot mark, as shown in the illustration here given so that it is practically level. To ascertain this level take a white or transparent glass bottle of any kind and fill it with water. There will be a little air space left which will manifest itself in the form of a bubble. Cork the bottle and lay it flatways on the stick. When a level has been reached the bubble will work its way along the top of the water to the center of the bottle. Having secured the level take the measurement of the distance from the stick to the ground at the end you are holding. Then multiply this measurement by 10 and you have a reasonably fair estimate of the grade. In the illustration, for instance, the distance from the ground to the stick is 18 inches. This multiplied by 10 gives 15 as the grade percentage, being ten times a foot and a half, or fifteen feet in the hundred.

Why Multiple "10" Is Used.

The reason for multiplying by 10 is that it is necessary to get the rise per 100 feet. As your stick, for convenience in handling, is only 10 feet long, the multiple 10 must be employed in order to get the grade, which, as explained further on, is always computed on the 100-feet basis.

It is peculiar fact that few people understand the real method of determining grades. As a rule they confuse angles with the percent of grades when in reality they are two radically different quantities. Thus a rise of 100 feet in 100 feet gives a percentage grade of 100, but the angle is one of 45 degrees. A glance at the illustration given will demonstrate this clearly. In referring to the inclination, the rise or fall, of a road surface, engineers invariably use the term "grade of — percent." They refer to a hill, for instance, as being of 5, 10, 15, 20 percent grade. This means that there is a vertical rise or fall of that number of feet to the 100 feet. In ascertaining this grade the engineer, by means of his instruments, measures off an exact 100 feet in a horizontal direction, regardless of the rise of the hill. He next measures off the exact height at the end of the 100 foot horizontal line, and the difference between this and "zero" or nothing at the starting point gives him the percentage of grade. Thus if the ground 100 feet from the starting point is 5 feet higher, the grade is a 5 percent one, and so on. While apparently an intricate one, the calculation is readily made by a process known to engineers as triangulation.

The next time you are on a motorcycle trip, and get into an argument with a brother rider over the grade of some hill you have climbed use the simple method first described and settle the controversy to the satisfaction of yourself and your companion.

Handy Shelter for Motorcycle.

Motorcycle owners, especially those who live in flats and lack proper housing or storage facilities for their machines, will find a valuable suggestion in the following:

Procure a large, heavy packing box of the proper dimensions. This should be not less than 4½ feet in height, 7½ feet in length, and 2½ feet wide. A box of this size and of good heavy material may usually be had for about $1.25. Take off one end and hinge it on to the box. Place the box so the hinged part of the end will be on the ground. This hinged end constitutes the door which, when open lies flat on the ground. Put a couple of layers of heavy roofing paper between the box and the ground. Then cover the top and the four sides with a layer of the same material. This makes it dry and damp proof. Add a hasp and padlock at the top of the cover and you have protection against prowling motorcycle thieves. When you wish to use the box open the door by laying it flat on the ground, and the machine may be readily pushed into, or pulled out of the box over it. This makes a cheap, handy and storm-proof housing.

Weight-Carrying Power of Tires.

Following is a handy table which gives the weight which may be carried on pneumatic-tired wheels according to the diameter of the tire and the air pressure at which it is inflated.

Diameter of Tire, Inches.	Maximum Weight on Wheels in Pounds.	Air Pressure Sq. Inches.
2½	225	50
3	350	60
3½	600	70
4	750	80
4½	1,000	90
5	1,000	90

HOW A NEGLECTED TIRE GROWS WORSE.

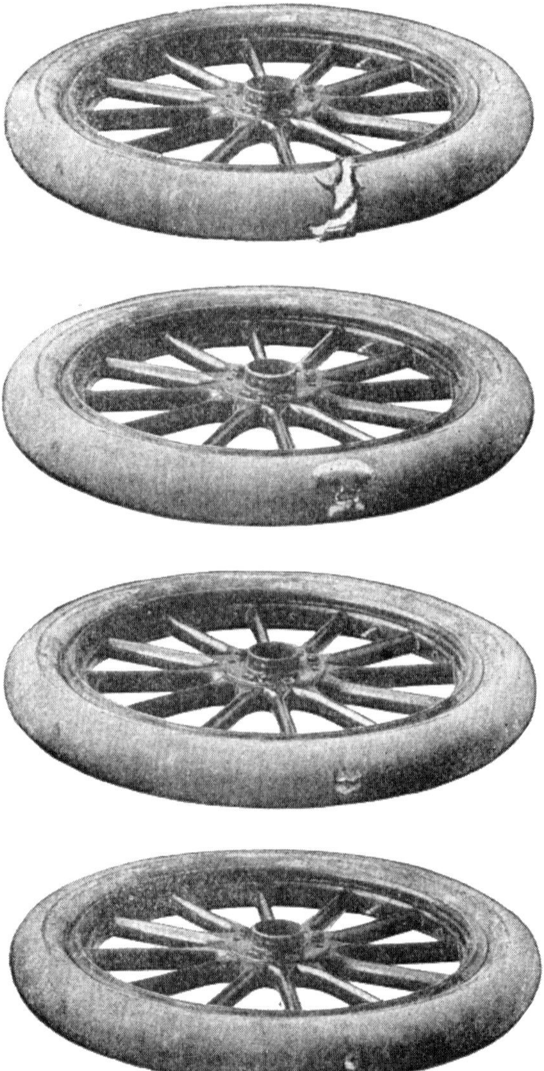

The First Break. It's Growing Larger. Bad Sand Pocket. Gone Beyond Repair.

REPAIRING AN INJURED TIRE BY THE SHALER METHOD.

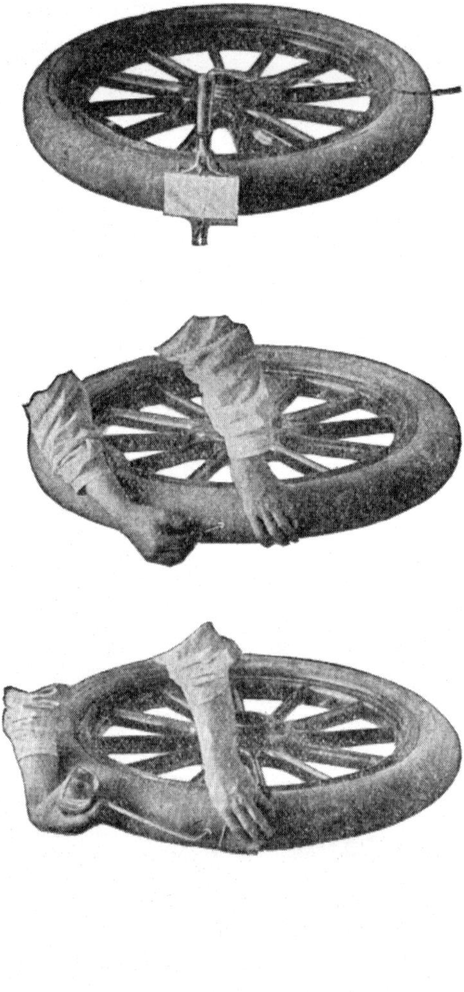

Cleaning the Cut. Filling Hole with Rubber. Using Shaler Vulcanizer.

RESULT—A NEW, SERVICEABLE TIRE.

CHAPTER XX.

BUYING A SECOND-HAND MOTORCYCLE.

As a general proposition, when a novice is tempted to buy a second-hand motorcycle, it is safe to say: "Don't do it!" Unless the owner has a good bona-fide reason for selling, it may be taken for granted that the machine is worn out, or there is something radically wrong with it. Taken as a rule it is safer and more satisfactory to buy a new machine, but, should conditions be such that it is a question of a second-hand machine or none, then a close careful inspection should be made. The nature of this inspection is well outlined by the Motorcyclist, as follows:

What to Do First.

The first thing to consider is whether the machine in question is worth purchasing at all. With the large selection offered one can afford to be particular; there are some faults which it is not worth while to attempt to remedy. The question of age resolves itself entirely into a question of make, for if the machine was a good one in the beginning it will be good much longer than one which was originally of a lower grade; but the matter of design is also important. A machine of short wheel-base, high saddle, high-hung engine, and short handle-bars will never be comfortable, and it would be wiser to wait for an opportunity to obtain one of a more modern build. Transmission must also be considered. The type is entirely a matter of personal preference; but block chains, chains too small for their work, a narrow flat belt with an idler, "V" pulleys not cut to twenty-eight degrees, or too small an engine pulley, can never be made to give efficient service. A cracked frame lug, cylinder, crank case, or piston, or a buckled wheel, also mean repairs that had better be left for the next fellow.

Examine the Engine Closely.

Not less important is the condition of the engine. This is a vital part of the machine and if it has been seriously

Idler Pulley.

injured the cycle will be part of the repair shop furniture for some time to come. Jack up the rear wheel and stand on the pedal with the valve dropped. A high compression engine in good condition will hold your weight almost indefinitely in this manner, and if the pedal does not fall for several seconds, the cylinder and rings may be considered to be in good condition. If it goes down rapidly the engine should be examined further. Have the cylinder taken off and run your finger carefully all over the inside surface. If you find any cuts, scratches, grooves, or rough spots, make a hurried exit. If, however, it is all bright and smooth, you may turn your attention to the rings. Examine carefully the surface which rubs against the cylinder. It should be all brightly polished, but there will probably be dark spots. If these spots are very frequent, you had better steer clear, but if only a few it simply means new rings. First, however, be sure that you can obtain new rings for that particular make and model, as it is an expensive job to have them specially made.

Is It Worth the Price.

Having satisfied yourself on all these points that the machine is really worth owning, the question arises: Is it worth the price asked? Of course, this can only be told by a knowledge of prevailing prices, though a careful examination will help you to compare it with other machines offered and enable you to chose the best.

While you have the cylinder off, glance at the bearings. Run off the belt (if the machine is so fitted) and try to move the pulley or engine sprocket up and down. A tiny amount of play is to be expected here, but if it is at all great, new main bushings will be needed, the cost of which must be added to the price you are to pay. The same is true of the crank and wrist pin bushings. The former can be tested by moving the connecting rod up and down while the pulley is held still, and the latter by holding the connecting rod with one hand and moving the piston up and down with the other.

Tires an Important Item.

Next to motor repairs the tires are probably the most expensive parts to renew. These are often overlooked in buying a second-hand mount, because they can be obtained anywhere, but it should be remembered that they cost a great deal. A flat tire means a puncture or rotten tube, although the owner may tell you that he just let it down "to keep it from stretching." Chains or belts are also large items and their condition should be noted. Do not forget to examine the sprockets, as if they are worn, it will be useless to replace the chains unless they are replaced also. Damaged pedalling gear, broken brake, dented tanks or rims, run down batteries, bent spokes, and broken mud-guard stays or control levers, are of less importance, as they are easily repaired, but they should all be looked for and taken into consideration in fixing the price. On the other hand, if the machine is fitted with a stand, lamp, horn, carrier, speedometer, or other extras, or spares, their value must be allowed for. Above all, be fair. Get what is due you, but, if you find the machine good, do not bicker about every little thing. Pay a reasonable price and get good wishes along with your motor.

CHAPTER XXI.

FOR PEOPLE WITH SPEED CRAZE.

There are some people who are speed crazy. Whenever they hear of some professional having made a mile in 46 seconds they are seized with a desire to do the same thing. They begin to despise their honest, stock machine capable of 35 or 40 miles an hour, and long to possess a regular "red devil." This may be done by making the following alterations, but we do not advise the changes, unless a man is going into the racing game, and then it will be cheaper to buy a racing motorcycle.

How to Convert a Motorcycle.

In the first place, says the Motorcyclist, we will change your gear because no motorcycle with a five or six to one gear can be speedy. About three to one is a fair gear for a racer. We will bore a series of holes around the cylinder so that a good part of the burnt gas will come out at the bottom of the stroke and relieve the back pressure and help clean the cylinder quickly of the burnt gases. Then we will put in a stiffer intake spring and maybe stiffen up the exhaust spring a little.

The exhaust port is far too small for a racer so we will cut away the entire front of the port. This will do no harm because a racer does not have an exhaust pipe muffler anyway and it will help the speed a lot. The carburetor also ought to be remodeled. After this we will bore the fly wheels full of holes and take off the bottom ring.

Then, we will, if our connecting rod is pretty thick, drill some holes in this or put in a new one.

Some Other Changes.

Most of those flanges on the cylinder might as well be taken off, and while we do that we will drill the ones left full of holes and also bore a few in the flanges of the head. We nearly forgot to raise the compression. Are we going to make a short distance racer of this or a ten to fifteen-mile racer? If a short distance one, we can cut off quite a bit from the lower end of the cylinder—just enough so the motor will not overheat in going a mile or two. You will not have any use for a throttle so we can do away with this and only use a free air opening. That "eats 'em alive." We will put on a good-sized pump and free oil opening.

And Then Some More.

The next alteration will be the replacing of the large, comfortable rubber pedals with light, rat-trap ones. The long, flexible grips will be traded for short, firm ones. If we can get the right kind of a coil we will use only two batteries and we will tap them on the frame to save the weight of the box. We ought to bore holes in the cranks also to lighten up a bit. The bars will have to be lowered for you will have to "laylow" to reduce wind resistance. Lowering the bars about 3 in. below the head will be about right. We will then shorten up the frame to about 52 inches or possibly 51 inches and place the saddle over the rear wheel. We will have to put on a small, light saddle because the three or four pounds difference in weight will help a whole lot. Then, you know, a very springy saddle does not give you as good control over your motorcycle as a more rigid one will.

What You Cannot Do.

Our racer is now complete. Isn't it a thing of beauty? I am sorry we did not get a lighter frame for this one is heavier than most racers and a few pounds make a lot of difference. You will have to get a pair of thick leggins and maybe will have to stuff some waste inside them where the heat from the exhaust ports come against them. If you do not you may get your calves blistered like

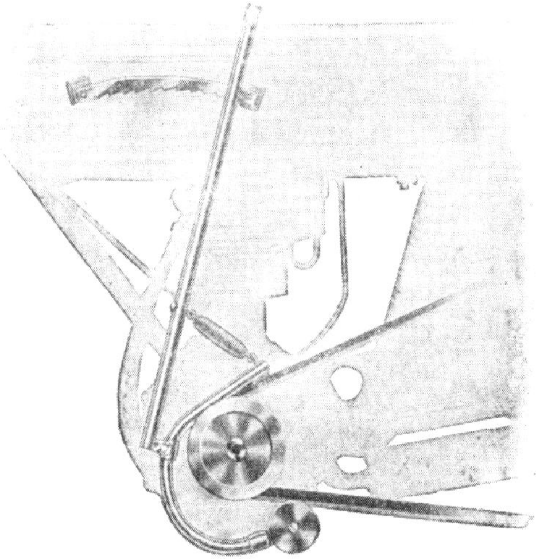

Tension of Idler, Pulley Released.

several racers have. Then you will want a good, big accident policy because several racers have had some narrow escapes. You also should have the local constable remove his eyes because this racer will not slow down like a nice motorcycle will. While he is removing them, have him remove all the grades in the city for you can not ride them with the high gear. Also have him sell all the surrounding country for you can not go very far with your machine.

And Then Throw It Away.

Get a rubber air cushion and tie it fast to your saddle and put a hinge in your back and springs in your wrists. Hire a couple of huskies to help push you every time you start. Establish gasolene storage stations every twenty-five miles, and then only ride in winter because your motor will overheat any place in America. Beg, borrow or steal lots of cash to buy new frames, tires, etc., with. Last, try your motorcycle on the road and you will find you cannot stay on top of it when you travel over 45 miles per hour and not over 25 on many roads and will find it pleasanter when riding about twenty per hour. If you are then still alive, trade the racer to one who thinks he doesn't know and buy a sweet running touring model from one who does know.

CHAPTER XXII.

"DON'TS" FOR MOTORCYCLISTS.

There are some things which are best impressed upon the mind in the form of "don'ts." Here are a few of them which the beginner in motorcycling will do well to keep before him:

Don't forget that gasolene is explosive—dangerously so—when exposed to the air. Consequently don't smoke while you are filling the gasolene tank, and don't light matches while looking into your tank in the dark to see if there is gasolene enough in it. Even some forms of lanterns—those with an air space around the bottom of the glass covering—are dangerous. The only real safety light is the portable electric.

Don't fill the gasolene tank to the limit; leave a little space for air. This will make it flow more readily by virtue of the pressure it exerts.

Don't allow a stranger to assist you in filling the tank; he may get the gasolene into the oil tank, or the oil into the gasolene. This has happened.

Don't forget to keep a new machine well oiled; over-oiled, if anything.

Don't ride with the muffler open if you can avoid it. The noise scares restive horses, and worries invalids and nervous people. It is the exhaust noise which leads many non-motorists to declare the motorcycle a nuisance. Don't forget that you can avoid a lot of this noise.

Don't ride past a motorcyclist who is stalled by the roadside. He may not need your help, and then again he may. You may be in the same fix yourself some day.

It will cost you nothing to slack up and inquire, "Can I help you?"

Don't travel at night without a light on your machine. Light it before it is really dark, and keep it going as long as you are on the road.

Don't forget that the color of the smoke issuing from the exhaust has a meaning; black smoke means too much gasolene; a bluish smoke means too much oil.

Don't forget that if you have engine trouble the first place to look for the cause is in the sparking system. Nine-tenths of the trouble of this nature will be located there.

Don't neglect to return spark plugs and such other accessories as you may have borrowed in time of emergency. You may wish to borrow again, and it is well to keep your credit good.

Don't overcrowd your motor, or let the engine race; it will put an unnecessary strain on it.

Don't ride at a racing speed, or turn corners at a rapid gait. It invites accidents which could be avoided, and creates a hostile feeling among the residents.

Don't forget to keep to the right when you are on the road. You have no business on the lefthand side; this belongs to "the other fellow."

Don't monkey with the carbureter, or magneto, or battery, except as a last resort. This same advice holds good as to the valve mechanism. Make sure of the gasolene supply and the sparking system first.

Don't forget to start slow and ride slow as much as you can. Sudden starts jar and overstrain the motor and are death on tires.

Don't use the brakes unnecessarily. Slow up gradually by shutting off the throttle. A sudden stop is worse even than a sudden start, and much more dangerous. It might cause the machine to act like a bucking broncho and unseat you.

Don't overlook the fact that pedestrians have the right of way at street crossings. Be especially careful in this respect where women and children are concerned.

Don't be too tenacious about your rights. The easiest way is the best. It is better to yield the right of way to some bull-headed teamster than to have your machine smashed by a heavy truck.

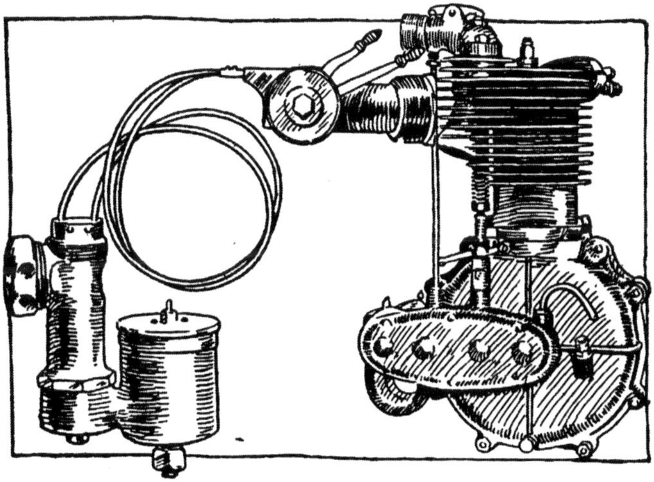

Magnet Carbureter Control and Motor.

Don't run away in case of accident. Stand by like a man, and be prompt to render such assistance as may be required. Don't get rattled. It is the rider with a clear head who comes out best.

Don't drink while on the road. Nine-tenths of such accidents as occur are due to befuddled brains.

Don't forget that, while your place is on the right side of the road, if you want to go ahead of some other rider moving in the same direction, you should, while still keeping on the right-hand side of the road, pass him on the left. This will prevent many a collision.

Don't neglect to shut off the gasolene and take out the switch plug when making a stop for any length of time. It will save fuel, and your machine, when you come back, is also likely to be where you left it.

Don't start on a long ride without due preparation, and be sure you have an emergency kit along. See list in chapter XVII.

Don't forget that in hill climbing your machine will do better work if you retard the spark and lower the speed. On level ground it will move better if the spark is advanced.

Don't forget that a motorcycle needs cleaning occasionally. It will not only look and run better, but will last longer.

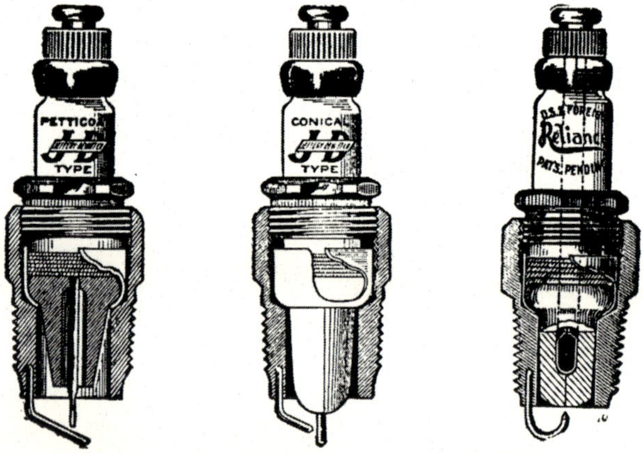

Three Forms of Jeffrey-Dewitt Spark Plugs.

Similar Books by Sportsman's Vintage Press

www.SportsmansVintagePress.com

ABC of the Motorcycle	by W. J. Jackman
Motorcycles in Competition	by Hank Wieand Bowman
Triumph Motorcycle Owner's Handbook	by A. St. J. Masters

Made in United States
North Haven, CT
04 November 2026

21772145R00136